Leckie
the education publisher
for Scotland

Higher
MATHS

Practice Question Book

Craig Lowther, Ken Nisbet

001/15082017

10 9 8 7 6

ISBN 9780008242329

Published by
Leckie
An imprint of HarperCollinsPublishers
Westerhill Road, Bishopbriggs
Glasgow
G64 2QT

HarperCollins Publishers
Macken House,
39/40 Mayor Street Upper,
Dublin 1
D01 C9W8
Ireland

T: 0844 576 8126   F: 0844 576 8131

leckiescotland@harpercollins.co.uk   www.leckiescotland.co.uk

Commissioning Editors: Clare Souza and Gillian Bowman
Managing Editor: Craig Balfour

Special thanks to
Jouve (layout)
Sarah Duxbury (cover design)
Project One Publishing Solutions (project management)
Philip Bradfield (answer check)
Ronald Gaffin (answer check)

A CIP Catalogue record for this book is available from the British Library.

Acknowledgements
Cover and p.1: 32 pixels / Shutterstock
p.82: J2R / Shutterstock
p.83: Flipser / Shutterstock

Whilst every effort has been made to trace the copyright holders, in cases where this has been unsuccessful, or if any have inadvertently been overlooked, the Publishers would gladly receive any information enabling them to rectify any error or omission at the first opportunity.

Printed in Great Britain by Ashford Colour Press Ltd.

FSC
www.fsc.org
MIX
Paper | Supporting responsible forestry
FSC™ C007454

This book contains FSC™ certified paper and other controlled sources to ensure responsible forest management.

For more information visit: www.harpercollins.co.uk/green

How to use this book

Welcome to Leckie's Higher Maths Practice Question Book. This book follows the structure of the Leckie Higher Maths Student Book, so is ideal to use alongside it. Questions have been written to provide practice for topics and concepts which have been identified as challenging for many students.

Examples

Examples with worked solutions provide support for particularly tricky concepts.

Use of calculators

Questions where you could use a calculator are marked with a icon.

Questions where you should **not** use a calculator are marked with a icon.

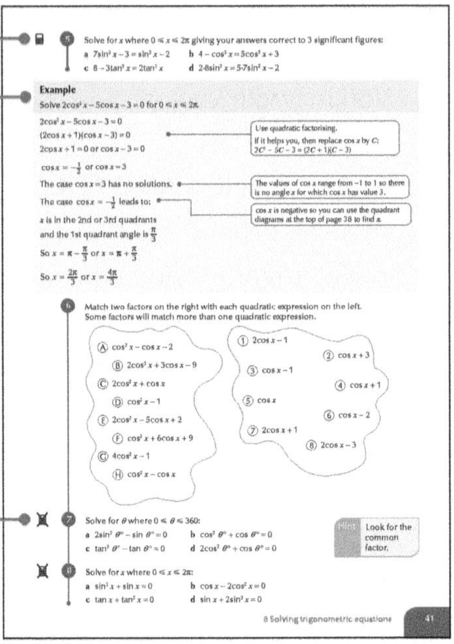

5 Solve for x where $0 \le x \le 2\pi$ giving your answers correct to 3 significant figures:

a $7\sin^2 x - 3 = \sin^2 x - 2$ b $4 - \cos^2 x = 5\cos^2 x + 3$

c $8 - 3\tan^2 x = 2\tan^2 x$ d $2{\cdot}8\sin^2 x = 5{\cdot}7\sin^2 x - 2$

Example

Solve $2\cos^2 x - 5\cos x - 3 = 0$ for $0 \le x \le 2\pi$.

$2\cos^2 x - 5\cos x - 3 = 0$

$(2\cos x + 1)(\cos x - 3) = 0$

$2\cos x + 1 = 0$ or $\cos x - 3 = 0$

Use quadratic factorising. If it helps you, then replace $\cos x$ by C: $2C^2 - 5C - 3 = (2C + 1)(C - 3)$

$\cos x = -\frac{1}{2}$ or $\cos x = 3$

The case $\cos x = 3$ has no solutions.

The values of $\cos x$ range from -1 to 1 so there is no angle x for which $\cos x$ has value 3.

The case $\cos x = -\frac{1}{2}$ leads to:

$\cos x$ is negative so you can use the quadrant diagrams at the top of page 38 to find x.

x is in the 2nd or 3rd quadrants and the 1st quadrant angle is $\frac{\pi}{3}$

So $x = \pi - \frac{\pi}{3}$ or $x = \pi + \frac{\pi}{3}$

So $x = \frac{2\pi}{3}$ or $x = \frac{4\pi}{3}$

6 Match two factors on the right with each quadratic expression on the left. Some factors will match more than one quadratic expression.

(A) $\cos^2 x - \cos x - 2$
(B) $2\cos^2 x + 3\cos x - 9$
(C) $2\cos^2 x + \cos x$
(D) $\cos^2 x - 1$
(E) $2\cos^2 x - 5\cos x + 2$
(F) $\cos^2 x + 6\cos x + 9$
(G) $4\cos^2 x - 1$
(H) $\cos^2 x - \cos x$

(1) $2\cos x - 1$
(2) $\cos x + 3$
(3) $\cos x - 1$
(4) $\cos x + 1$
(5) $\cos x$
(6) $\cos x - 2$
(7) $2\cos x + 1$
(8) $2\cos x - 3$

7 Solve for θ where $0 \le \theta \le 360$:

a $2\sin^2 \theta° - \sin \theta° = 0$ b $\cos^2 \theta° + \cos \theta° = 0$

c $\tan^2 \theta° - \tan \theta° = 0$ d $2\cos^2 \theta° + \cos \theta° = 0$

Hint: Look for the common factor.

8 Solve for x where $0 \le x \le 2\pi$:

a $\sin^2 x + \sin x = 0$ b $\cos x - 2\cos^2 x = 0$

c $\tan x + \tan^2 x = 0$ d $\sin x + 2\sin^2 x = 0$

8 Solving trigonometric equations 41

Hints

Where appropriate, hints are provided to help give extra guidance and support.

Answers

Check your own work. The answers are provided online at:

https://collins.co.uk/pages/scottish-curriculum-free-resources

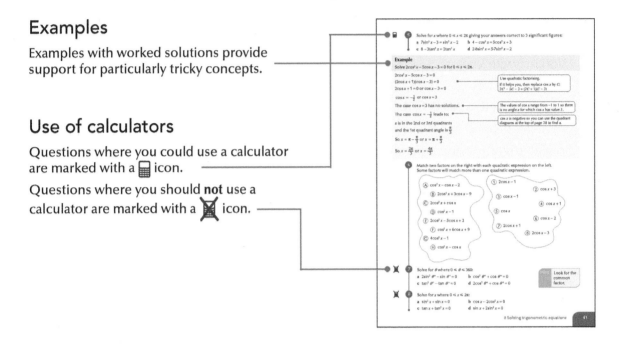

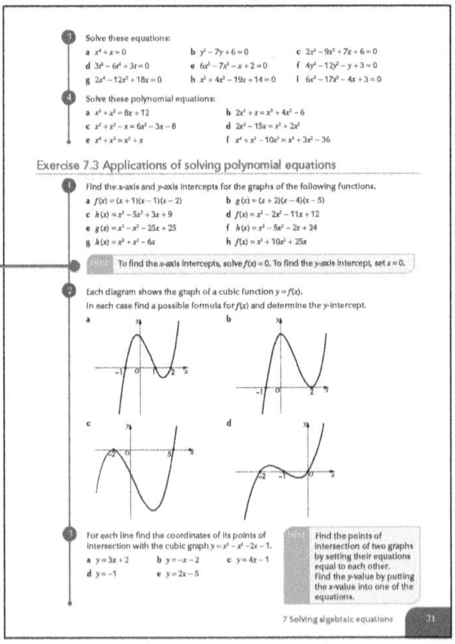

3 Solve these equations:

a $x^4 + x = 0$ b $y^3 - 7y + 6 = 0$ c $2x^3 - 9x^2 + 7x + 6 = 0$

d $3t^3 - 6t^2 + 3t = 0$ e $6x^3 - 7x^2 - x + 2 = 0$ f $4y^3 - 12y^2 - y + 3 = 0$

g $2x^4 - 12x^3 + 18x = 0$ h $x^3 + 4x^2 - 19x + 14 = 0$ i $6x^3 - 17x^2 - 4x + 3 = 0$

4 Solve these polynomial equations:

a $x^3 + x^2 = 8x + 12$ b $2x^3 + x = x^3 + 4x^2 - 6$

c $x^3 + x^2 - x = 6x^2 - 3x - 8$ d $2x^3 - 15x = x^3 + 2x^2$

e $x^4 + x^3 = x^2 + x$ f $x^4 + x^3 - 10x^2 = x^3 + 3x^2 - 36$

Exercise 7.3 Applications of solving polynomial equations

1 Find the x-axis and y-axis intercepts for the graphs of the following functions.

a $f(x) = (x + 1)(x - 1)(x - 2)$ b $g(x) = (x + 2)(x - 4)(x - 5)$

c $h(x) = x^3 - 5x^2 + 3x + 9$ d $f(x) = x^3 - 2x^2 - 11x + 12$

e $g(x) = x^3 - x^2 - 25x + 25$ f $h(x) = x^3 - 5x^2 - 2x + 24$

g $h(x) = x^3 + x^2 - 6x$ h $f(x) = x^3 + 10x^2 + 25x$

Hint: To find the x-axis intercepts, solve $f(x) = 0$. To find the y-axis intercept, set $x = 0$.

2 Each diagram shows the graph of a cubic function $y = f(x)$.
In each case find a possible formula for $f(x)$ and determine the y-intercept.

a b c d

3 For each line find the coordinates of its points of intersection with the cubic graph $y = x^3 - x^2 - 2x - 1$.

a $y = 3x + 2$ b $y = -x - 2$ c $y = 4x - 1$

d $y = -1$ e $y = 2x - 5$

Hint: Find the points of intersection of two graphs by setting their equations equal to each other. Find the y-value by putting the x-value into one of the equations.

7 Solving algebraic equations 31

1 Manipulating algebraic expressions

Exercise 1.1 Logarithms and exponentials

1 Express in logarithmic form:

a $8 = 2^3$ **b** $9 = 3^2$ **c** $81 = 3^4$ **d** $125 = 5^3$

e $a = 4^2$ **f** $p = q^r$ **g** $x = e^5$ **h** $G = 10^y$

2 Express in exponential form:

a $4 = \log_5 625$ **b** $3 = \log_{10} 1000$ **c** $6 = \log_2 64$ **d** $a = \log_b c$

e $x = \log_5 y$ **f** $y = \log_5 x$ **g** $x = \log_y 5$ **h** $y = \log_x 5$

3 Evaluate:

a $\log_2 8$ **b** $\log_{10} 100$ **c** $\log_3 3$ **d** $\log_2 \frac{1}{2}$

e $\log_7 49$ **f** $\log_5 1$ **g** $\log_{10} 0{\cdot}1$ **h** $\log_2 \sqrt{2}$

> **Hint** To evaluate $\log_a b$ ask: 'To what power do I raise a to get b?'

4 Simplify and then evaluate:

a $\log_6 2 + \log_6 3$ **b** $\log_2 36 - \log_2 9$ **c** $\log_{10} 4 + \log_{10} 25$

d $\log_3 54 - \log_3 2$ **e** $\log_{10} 2 - \log_{10} 20$ **f** $\log_7 1 + \log_7 7$

g $\log_{12} 4 + \log_{12} 3$ **h** $\log_6 8 + \log_6 9 - \log_6 2$

i $\log_2 8 + \log_2 6 - \log_2 3$ **j** $\log_3 8 - \log_3 4 - \log_3 6$

> **Reminder**
>
> $\log_a(xy) = \log_a x + \log_a y$
>
> $\log_a\left(\frac{x}{y}\right) = \log_a x - \log_a y$
>
> $\log_a(x^n) = n\log_a x$

5 Simplify:

a $\log_2 100 - 2\log_2 5$ **b** $\log_{10} 8 + \log_{10} 5 - \log_{10} 4$

c $\frac{1}{2}\log_6 9 + \log_6 2$ **d** $\log_4 \sqrt{80} - \log_4 \sqrt{5}$

6 Calculate giving your answers correct to 2 decimal places:

a $\log_e 3{\cdot}8$ **b** $\log_{10} 3{\cdot}8$ **c** $e^{1{\cdot}4}$ **d** $10^{1{\cdot}4}$

e $e^{-0{\cdot}7}$ **f** $\log_e 0{\cdot}42$ **g** $\log_{10} 0{\cdot}97$ **h** $\log_e 10^{1{\cdot}5}$

> **Hint** To calculate $\log_{10}$ use **log** button
>
> To calculate $\log_e$ use **ln** button

Exercise 1.2 Solving logarithmic and exponential equations

Example

Solve: **a** $\log_e x = 5$ **b** $\log_{10} x = 2{\cdot}9$ **c** $e^x = 4{\cdot}5$ **d** $10^x = 2$

a $\log_e x = 5 \Rightarrow x = e^5 \approx 148{\cdot}4$ (using e^x)

b $\log_{10} x = 2{\cdot}9 \Rightarrow x = 10^{2{\cdot}9} \approx 794{\cdot}3$ (using 10^x)

c $e^x = 4{\cdot}5 \Rightarrow x = \ln 4{\cdot}5 \approx 1{\cdot}50$ (using **ln**)

d $10^x = 2 \Rightarrow x = \log_{10} 2 \approx 0{\cdot}301$ (using **log**)

> Change from one of these to the other:
>
> 'power statement': $a = b^c$
>
> 'log statement': $\log_b a = c$

1 Change the following 'power statements' to the corresponding 'log statements'.

a $92 = 10^x$ b $4 = e^{-y}$ c $e^{2t} = 14{\cdot}1$ d $10^{0.5t} = y$

e $0{\cdot}002 = e^{1.5t}$ f $e^{-5t} = B_3$ g $e^{-2x} = 0{\cdot}5$ h $A_1 = 10^{x+t}$

2 Change these 'log statements' to 'power statements':

a $\log_2 x = 3$ b $\log_5 10 = x$ c $4 = \log_{10} t$

d $\log_e A = B$ e $\log_m V_0 = k$ f $\log_b (2A + 1) = T$

3 Solve for x giving your answer correct to 3 significant figures:

a $\log_e x = 1{\cdot}2$ b $\log_e x = 6$ c $\log_e x = 3{\cdot}25$ d $\log_e x = 0{\cdot}2$

e $\log_{10} x = 1{\cdot}4$ f $\log_{10} x = 2{\cdot}3$ g $\log_{10} x = 0{\cdot}03$ h $\log_{10} x = 3{\cdot}305$

i $e^x = 7$ j $e^x = 0{\cdot}2$ k $e^x = 45$ l $e^x = 100$

m $10^x = 0{\cdot}5$ n $10^x = 67$ o $10^x = 2017$ p $10^x = 0{\cdot}002$

4 Solve for x correct to 3 significant figures:

a $\log_e x = 2$ b $\log_{10} x = 3{\cdot}1$ c $e^x = 3$ d $10^x = 53$

e $\log_3 x = 0{\cdot}03$ f $\log_5 x = 0{\cdot}32$ g $e^x = 0{\cdot}03$ h $10^x = 0{\cdot}15$

i $\log_2 x = 5$ j $\log_7 x = 1{\cdot}4$ k $\log_9 x = 0{\cdot}8$ l $\log_3 x = 0{\cdot}1$

m $27 = 10^x$ n $6{\cdot}7 = \log_e x$ o $4{\cdot}1 = \log_{10} x$ p $32 = e^x$

Hint Check your calculator for one of these:

$x^{\blacksquare}$ ^ x^y y^x

That's the button you need to calculate $3^{0.03}$

5 a How do you write the year 2018 as a power of 10 correct to 3 significant figures?

b Now write 2018 as a power of e correct to 3 significant figures.

6 Solve for x:

a $\log_2 (x + 2) + \log_2 4 = 4$ b $\log_5 (2x + 3) - \log_5 x = 1$

c $\log_6 (x + 1) + \log_6 x = 1$; $(x > 0)$ d $\log_3 (x - 3) + \log_3 x = \log_3 18$; $(x > 0)$

7 Evaluate (correct to 3 significant figures):

a $\log_2 3$ b $\log_5 6$ c $\log_8 40$

d $\log_4 0{\cdot}2$ e $\log_7 \left(\frac{1}{2}\right)$ f $\log_3 \sqrt{2}$

Hint Write $x = \log_2 3$. Rewrite as a 'power statement', then take the logs of both sides.

8 Solve (correct to 3 significant figures):

a $2^x = 1000$ b $3^x = 0{\cdot}1$ c $7^x = \frac{3}{2}$

d $20^x = 2$ e $5^{-x} = 0{\cdot}01$ f $4^{-0.1x} = 0{\cdot}328$

2 Manipulating trigonometric expressions

Exercise 2.1 Degrees, radians and exact values

Reminder

For $\frac{\pi}{4}$ or 45° use half a square of side 1:

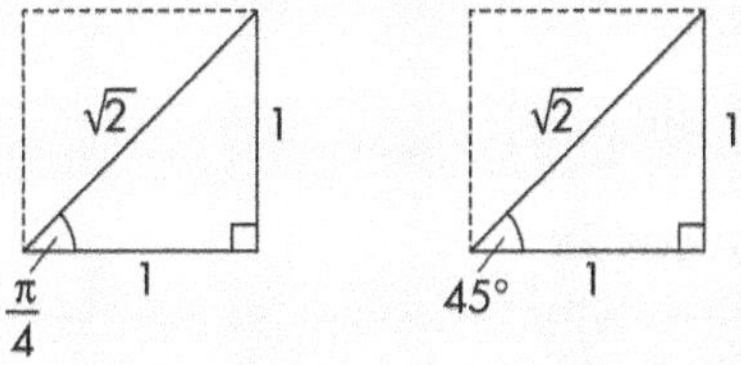

Now use 'SOHCAHTOA' to find **exact** values.

For $\frac{\pi}{6}, \frac{\pi}{3}$ or 30°, 60° use half an equilateral triangle of side 2:

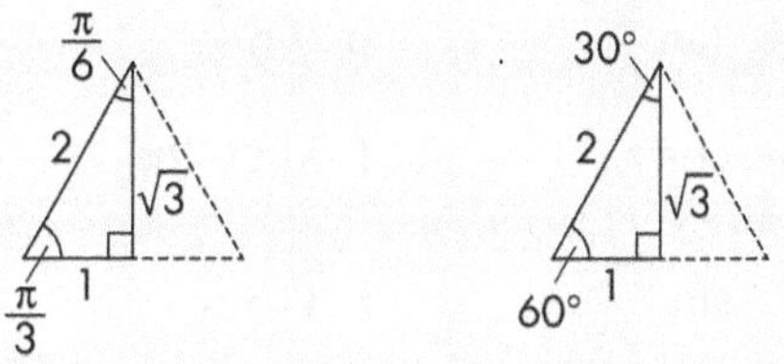

1 Find the exact value of:

a sin 30° **b** cos 45° **c** tan 60° **d** tan 120°
e sin 135° **f** cos 405° **g** sin (−30)° **h** tan (−60)°
i cos (−45)° **j** cos (−300)° **k** sin 570° **l** cos 225°
m tan 330° **n** tan 495° **o** cos (−150)° **p** sin 420°
q sin (−150)° **r** tan 390° **s** sin (−270)° **t** cos (−450)°

2 In each case find the exact value of sin A, cos A and tan A.

a

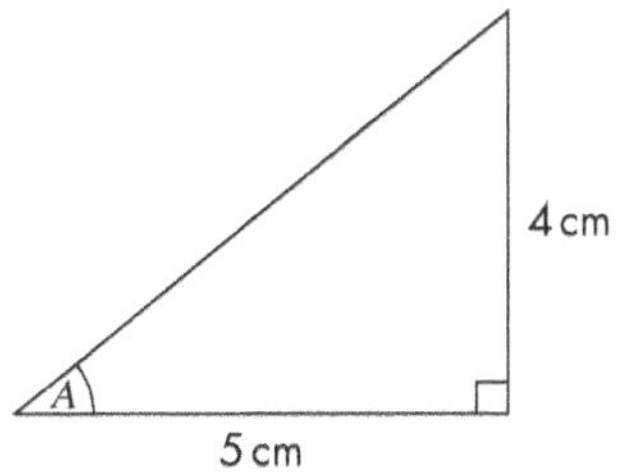

b

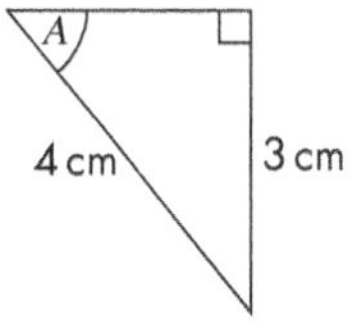

c

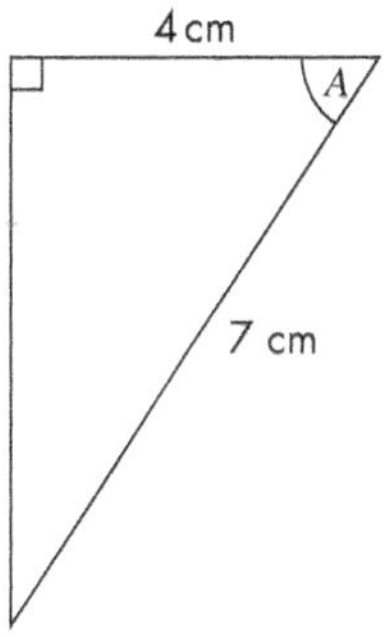

3 Convert these angles to radians:

a 90° **b** 210° **c** 330° **d** 360°
e 60° **f** −120° **g** −45° **h** 180°
i 390° **j** 420° **k** 270° **l** 45°
m −135° **n** 225° **o** −150° **p** −240°

4 Simplify to give a single angle in radians:

a $\pi + \frac{\pi}{4}$ **b** $2\pi - \frac{\pi}{3}$ **c** $\pi - \frac{\pi}{6}$
d $2\pi + \frac{\pi}{6}$ **e** $\pi - \frac{\pi}{3}$ **f** $\pi + \frac{\pi}{6}$
g $2\pi - \frac{\pi}{4}$ **h** $\pi + \frac{\pi}{3}$ **i** $\pi - \frac{\pi}{4}$
j $2\pi - \frac{\pi}{6}$ **k** $2\pi + \frac{\pi}{4}$ **l** $2\pi + \frac{\pi}{3}$

Hint $2\pi - \frac{\pi}{3}$ can be thought of as: '6 lots of $\frac{\pi}{3}$ minus 1 lot of $\frac{\pi}{3}$'.

5 Convert these angles to degrees:

a $\frac{\pi}{6}$ b $\frac{2\pi}{3}$ c $\frac{7\pi}{4}$

d $-\frac{\pi}{2}$ e $-\frac{5\pi}{4}$ f $\frac{4\pi}{3}$

g $\frac{11\pi}{6}$ h $\frac{5\pi}{3}$ i $\frac{\pi}{3}$

j π k $-\frac{5\pi}{3}$ l $\frac{5\pi}{4}$

Hint $\frac{7\pi}{3}$ can be thought of as '7 lots of $\frac{\pi}{3}$' so '7 lots of 60°'.

6 Find the exact value of:

a $\tan\frac{\pi}{6}$ b $\sin\frac{\pi}{3}$ c $\tan\frac{7\pi}{6}$ d $\sin\left(-\frac{\pi}{3}\right)$

e $\cos\pi$ f $\sin(-2\pi)$ g $\cos\frac{5\pi}{3}$ h $\cos 2\pi$

i $\sin\frac{2\pi}{3}$ j $\cos\frac{\pi}{2}$ k $\sin\frac{5\pi}{4}$ l $\tan\frac{4\pi}{3}$

Exercise 2.2 The addition formulae

Example

Show that the exact value of $\sin(x+y)°$ is $\frac{378+60\sqrt{13}}{609}$

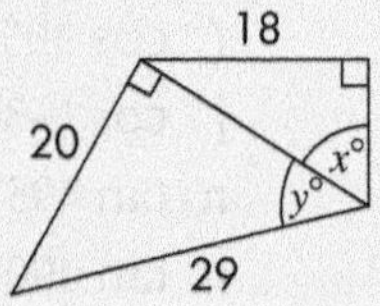

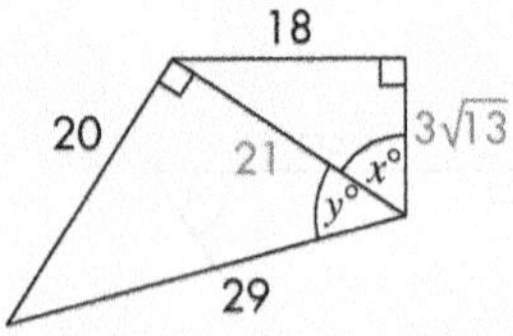

$\sqrt{29^2-20^2} = \sqrt{441} = 21$ — Use Pythagoras' theorem to find the two missing lengths.

$\sqrt{21^2-18^2} = \sqrt{117} = \sqrt{9\times 13} = 3\sqrt{13}$

$\sin(x+y)° = \sin x°\cos y° + \cos x°\sin y°$ — Use SOHCAHTOA in the right-angled triangles to find the correct ratio values.

$= \frac{18}{21}\times\frac{21}{29}+\frac{3\sqrt{13}}{21}\times\frac{20}{29}$

$= \frac{18\times 21}{21\times 29}+\frac{3\sqrt{13}\times 20}{21\times 29}$ — Denominators are the same so you can add the numerators.

$= \frac{18\times 21+60\sqrt{13}}{21\times 29}$

$= \frac{378+60\sqrt{13}}{609}$

1 Expand these and simplify if possible. (Don't look at the formulae. Try to remember them.)

a $\cos(C-D)°$ b $\sin(\theta+\phi)$ c $\cos\left(2x+\frac{\pi}{6}\right)$ d $\sin(x-\pi)$

e $\cos(180-x)°$ f $\cos(45+3x)°$ g $\sin\left(\frac{x}{2}-45\right)°$ h $\sin\left(\frac{\pi}{3}+\frac{x}{2}\right)$

2 Find the exact value of:

a $\sin A$ b $\cos A$

c $\sin B$ d $\cos B$

e $\sin(A-B)$ f $\sin(A+B)$

g $\cos(A-B)$ h $\cos(A+B)$

i $\tan(A-B)$ j $\tan(A+B)$

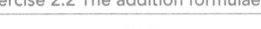

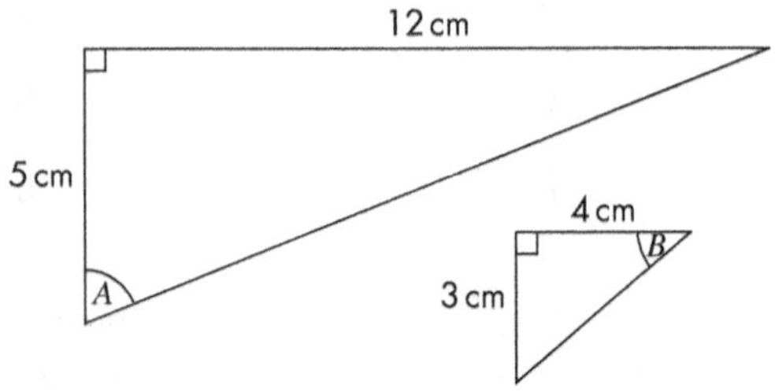

3 **a** Show that the exact value of $\sin(x+y)°$ is $\dfrac{10\sqrt{14}+60}{117}$

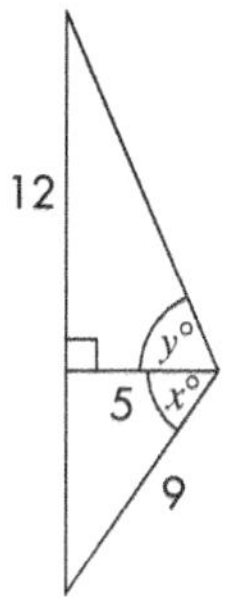

b Show that the exact value of $\cos(x+y)°$ is $\dfrac{156-40\sqrt{3}}{247}$

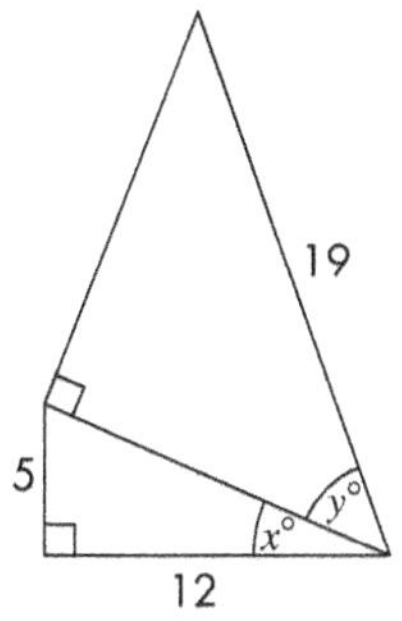

c Show that $\cos(x+y)° = \dfrac{1}{\sqrt{5}}$

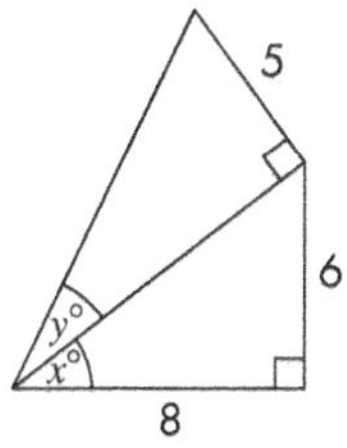

d $AB = 9$ cm, $AC = 11$ cm and $BD = 12$ cm.

If $\angle BAD = x°$ and $\angle BAC = y°$ then show that

$\sin(x-y)° = \dfrac{108-18\sqrt{10}}{165}$

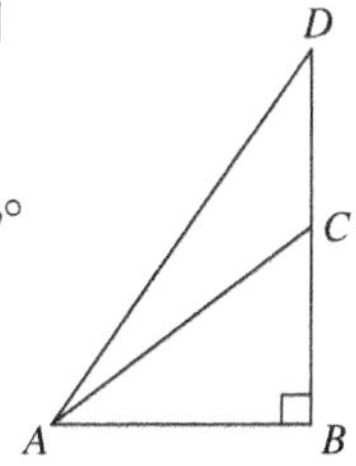

4 Expand and calculate the exact values.

a $\cos(45-30)°$ **b** $\sin\left(\frac{\pi}{6}-\frac{\pi}{4}\right)$ **c** $\cos(30+45)°$ **d** $\cos\left(\frac{\pi}{3}+\frac{\pi}{4}\right)$

5 Use the addition formulae to calculate the exact values.

a $\sin 15°$ **b** $\sin 105°$ **c** $\tan 15°$

d $\cos 75°$ **e** $\tan 75°$ **f** $\tan 105°$

6 Show using the addition formula that $\sin\left(\frac{\pi}{2}+\frac{\pi}{3}\right) = \frac{1}{2}$

7 By writing $\frac{7\pi}{4}$ as $\frac{3\pi}{2}+\frac{\pi}{4}$ show that $\cos\frac{7\pi}{4} = \frac{1}{\sqrt{2}}$

Exercise 2.3 The double angle formulae

Example

If $\tan\alpha = \dfrac{\sqrt{7}}{2}$ find the exact value of $\cos 2\alpha$ where $0 < \alpha < \dfrac{\pi}{2}$

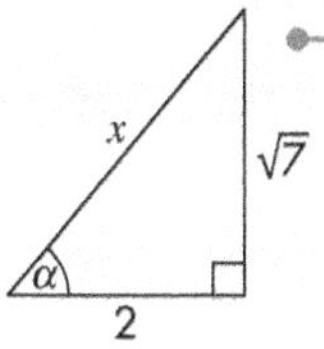

$x^2 = 2^2 + \left(\sqrt{7}\right)^2 = 4 + 7 = 11$

So $x = \sqrt{11}$

$\cos 2\alpha = 2\cos^2\alpha - 1$

$= 2\times\left(\dfrac{2}{\sqrt{11}}\right)^2 - 1$

$= 2\times\dfrac{4}{11} - 1 = \dfrac{8}{11} - 1 = \dfrac{8}{11} - \dfrac{11}{11} = -\dfrac{3}{11}$

Here is a right-angled triangle for which $\tan\alpha = \dfrac{\sqrt{7}}{2}$

Use Pythagoras' theorem to find the length of the third side.

Here you could also have used $\cos^2\alpha - \sin^2\alpha$ or $1 - 2\sin^2\alpha$.

Finally substitute the correct ratio values from the triangle and then simplify.

1 In each case $0 < \alpha < \frac{\pi}{2}$. Draw a right-angled triangle to find the required exact value.

a $\sin\alpha = \frac{1}{\sqrt{5}}$. Find $\cos\alpha$.

b $\tan\alpha = \frac{\sqrt{3}}{4}$. Find $\sin\alpha$.

c $\cos\alpha = \frac{1}{8}$. Find $\tan\alpha$.

d $\sin\alpha = \frac{\sqrt{3}}{\sqrt{5}}$. Find $\tan\alpha$.

e $\tan\alpha = \frac{\sqrt{7}}{\sqrt{3}}$. Find $\cos\alpha$.

2 If $0 < \alpha < \frac{\pi}{2}$, find the exact value of $\sin 2\alpha$ given that:

> **Hint** $\sin 2\alpha = 2\sin\alpha\cos\alpha$

a $\tan\alpha = \frac{1}{3}$ **b** $\cos\alpha = \frac{2}{\sqrt{7}}$ **c** $\sin\alpha = \frac{\sqrt{2}}{5}$ **d** $\cos\alpha = \frac{\sqrt{3}}{\sqrt{5}}$

3 If $0 < \theta < \frac{\pi}{2}$, find the exact value of $\cos 2\theta$ given that:

a $\cos\theta = \frac{3}{4}$ **b** $\sin\theta = \frac{5}{\sqrt{29}}$ **c** $\cos\theta = \frac{\sqrt{2}}{\sqrt{7}}$ **d** $\tan\theta = \frac{4}{\sqrt{33}}$

4 Use the double angle formula to rewrite:

a $\sin 2B°$ **b** $\sin\theta$ **c** $\sin\alpha$

d $\sin 4x°$ **e** $\sin D°$ **f** $\sin 3x°$

g $\sin\left(\frac{x}{2}\right)°$ **h** $\sin\left(\frac{2x}{3}\right)°$ **i** $\sin\frac{\pi}{3}$

5 For each diagram find the exact value of: **i** $\sin\left(\frac{x}{2}\right)°$ **ii** $\cos\left(\frac{x}{2}\right)°$

a

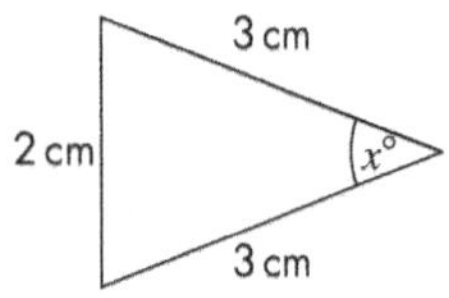

b

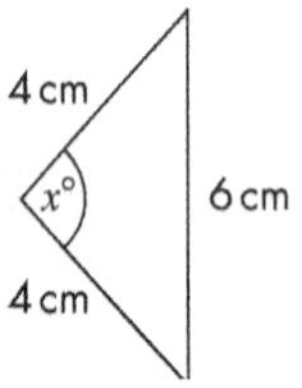

c

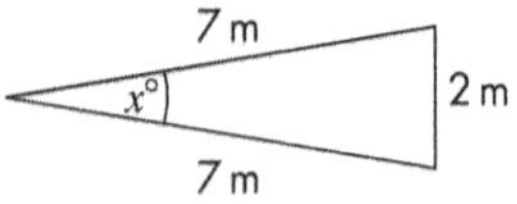

d

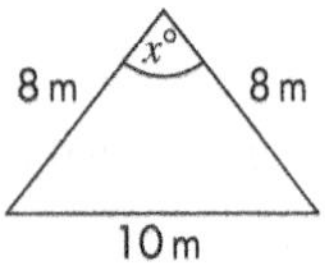

e

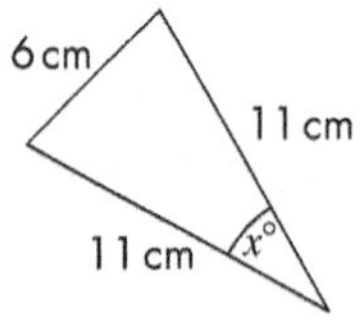

> **Hint** Draw in the altitude. That is the axis of symmetry for the triangle.

6 In each case find the exact value of $\sin x°$.

a

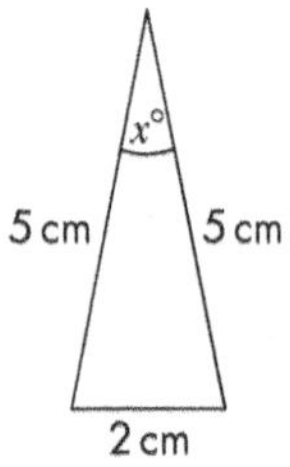

b

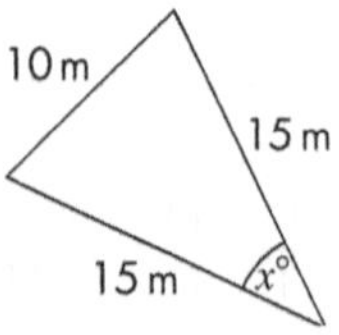

c

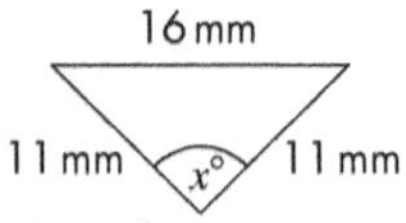

> **Hint** Rewrite $\sin x°$ using the double angle formula.

7 A billboard is supported by a wooden support wedge as shown in the diagram on the left. The wedge has dimensions as shown in the diagram on the right.

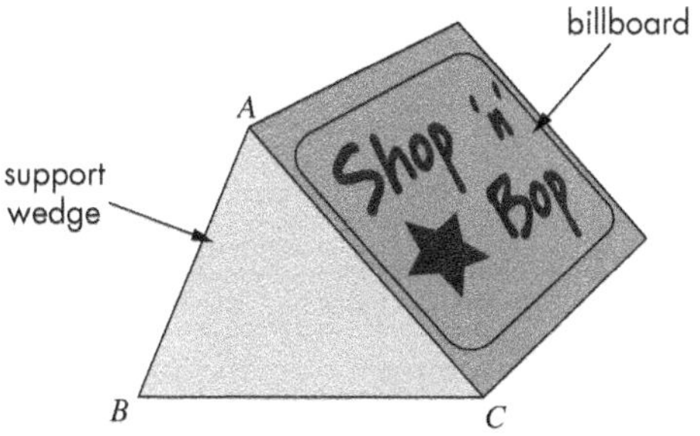

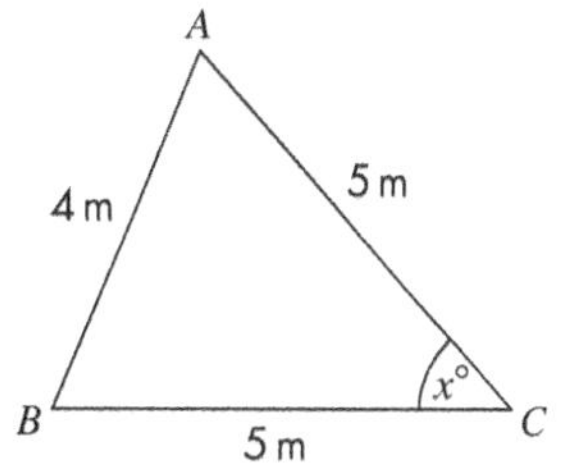

Find the exact value of sin $x°$ where $x°$ is the angle at which the billboard is inclined to the horizontal.

Exercise 2.4 Trig identities

Reminder

You will find these formulae useful in this exercise:

$\sin(A+B) = \sin A\cos B + \cos A\sin B$

$\sin(A-B) = \sin A\cos B - \cos A\sin B$

$\cos(A+B) = \cos A\cos B - \sin A\sin B$

$\cos(A-B) = \cos A\cos B + \sin A\sin B$

$\sin 2A = 2\sin A\cos A$

$\cos 2A = 2\cos^2 A - 1$

or $\cos 2A = \cos^2 A - \sin^2 A$

or $\cos 2A = 1 - 2\sin^2 A$

$\sin^2 A + \cos^2 A = 1$

$\tan A = \dfrac{\sin A}{\cos A}$

1 Show that:

a $\sin\left(\frac{\pi}{2} + A\right) = \cos(2\pi - A)$

b $\cos(180 - A)° = \sin(270 - A)°$

c $\cos\left(\frac{3\pi}{2} - P\right) = \sin(2\pi - P)$

d $\cos(270 - \theta)° + \cos(270 + \theta)° = 0$

e $\sin\left(\frac{\pi}{2} + A\right) - \cos A = \cos\left(\frac{\pi}{2} + A\right) + \sin A$

f $\sin(30 - x)° + \cos(60 + x)° = \cos x° - \sqrt{3}\sin x°$

g $\cos(60 - x)° - \frac{\sqrt{3}}{2}\sin x° = \frac{1}{2}\cos x°$

h $\sin\left(\frac{\pi}{4} + B\right) - \cos\left(\frac{\pi}{4} - B\right) = 0$

2 Prove the following identities.

a $(\cos x - \sin x)^2 = 1 - \sin 2x$

b $(\cos\theta + \sin\theta)(\cos\theta - \sin\theta) = \cos 2\theta$

c $(2\cos t - 1)(2\cos t + 1) = 2\cos 2t + 1$

d $\dfrac{\sin 2x}{1 + \cos 2x} = \tan x$

e $\dfrac{1 - \cos 2\phi}{1 + \cos 2\phi} = \tan^2\phi$

Exercise 2.5 The wave function

Example

Solve the simultaneous equations:

$k\sin\alpha° = 2$

$k\cos\alpha° = -5$ where $k > 0$ and $0 \leqslant \alpha < 360$.

$\frac{k\sin\alpha°}{k\cos\alpha°} = \frac{2}{-5}$ — Divide the two sides of the equations and use $\frac{\sin\alpha°}{\cos\alpha°} = \tan\alpha°$

$\tan\alpha° = -\frac{2}{5} = -0{\cdot}4$

Use $\tan^{-1} 0{\cdot}4$ for the 1st quadrant angle $21{\cdot}8$.

$\alpha = 180 - 21{\cdot}8 = 158{\cdot}2$ — α is in 2nd quadrant: $\sin\alpha° > 0$ and $\cos\alpha° < 0$.

$k^2 = 2^2 + (-5)^2 = 29$ — This comes from squaring and adding both sides of the two equations.

so $k = \sqrt{29}$

1 Find x for each pair of equations, where $0 \leqslant x \leqslant 360$ and $k > 0$.

a $k\sin x° = 1$, $k\cos x° = 2$

b $k\sin x° = 3$, $k\cos x° = 2$

c $k\sin x° = \sqrt{3}$, $k\cos x° = 1$

d $k\sin x° = 3$, $k\cos x° = 3$

2 Find k for each pair of equations ($k > 0$).

a $k\sin x° = 2$, $k\cos x° = 1$

b $k\sin x° = 5$, $k\cos x° = 3$

c $k\sin x° = \sqrt{3}$, $k\cos x° = 2$

d $k\sin x° = 1$, $k\cos x° = \sqrt{2}$

3 Solve these simultaneous equations where $k > 0$ and $0 \leqslant \alpha \leqslant 360$.

a $k\sin\alpha° = 4$, $k\cos\alpha° = 3$

b $k\sin\alpha° = \sqrt{7}$, $k\cos\alpha° = 1$

c $k\sin\alpha° = 5$, $k\cos\alpha° = 5$

d $k\sin\alpha° = 8$, $k\cos\alpha° = 3$

e $k\sin\alpha° = \sqrt{3}$, $k\cos\alpha° = \sqrt{5}$

f $k\sin\alpha° = \sqrt{10}$, $k\cos\alpha° = \sqrt{5}$

4 In each case express $f(x)$ in the given form with $k > 0$ and $0 \leqslant \alpha \leqslant 360$.

a $f(x) = 3\cos x° + 4\sin x°$ in the form $k\cos(x - \alpha)°$
[use $\cos(x - \alpha)° = \cos x° \cos\alpha° + \sin x° \sin\alpha°$]

b $f(x) = 5\sin x° + 12\cos x°$ in the form $k\sin(x + \alpha)°$
[use $\sin(x + \alpha)° = \sin x° \cos\alpha° + \cos x° \sin\alpha°$]

c $f(x) = \sin x° - 2\cos x°$ in the form $k\sin(x - \alpha)°$

d $f(x) = 2\cos x° - 5\sin x°$ in the form $k\cos(x + \alpha)°$

e $f(x) = \sqrt{3}\cos x° + \sin x°$ in the form $k\cos(x - \alpha)°$

f $f(x) = \sqrt{2}\sin x° - \sqrt{2}\cos x°$ in the form $k\sin(x - \alpha)°$

3 Identifying and sketching related functions

Exercise 3.1 Graphs of related functions

Examples of transformations and their related graphs.

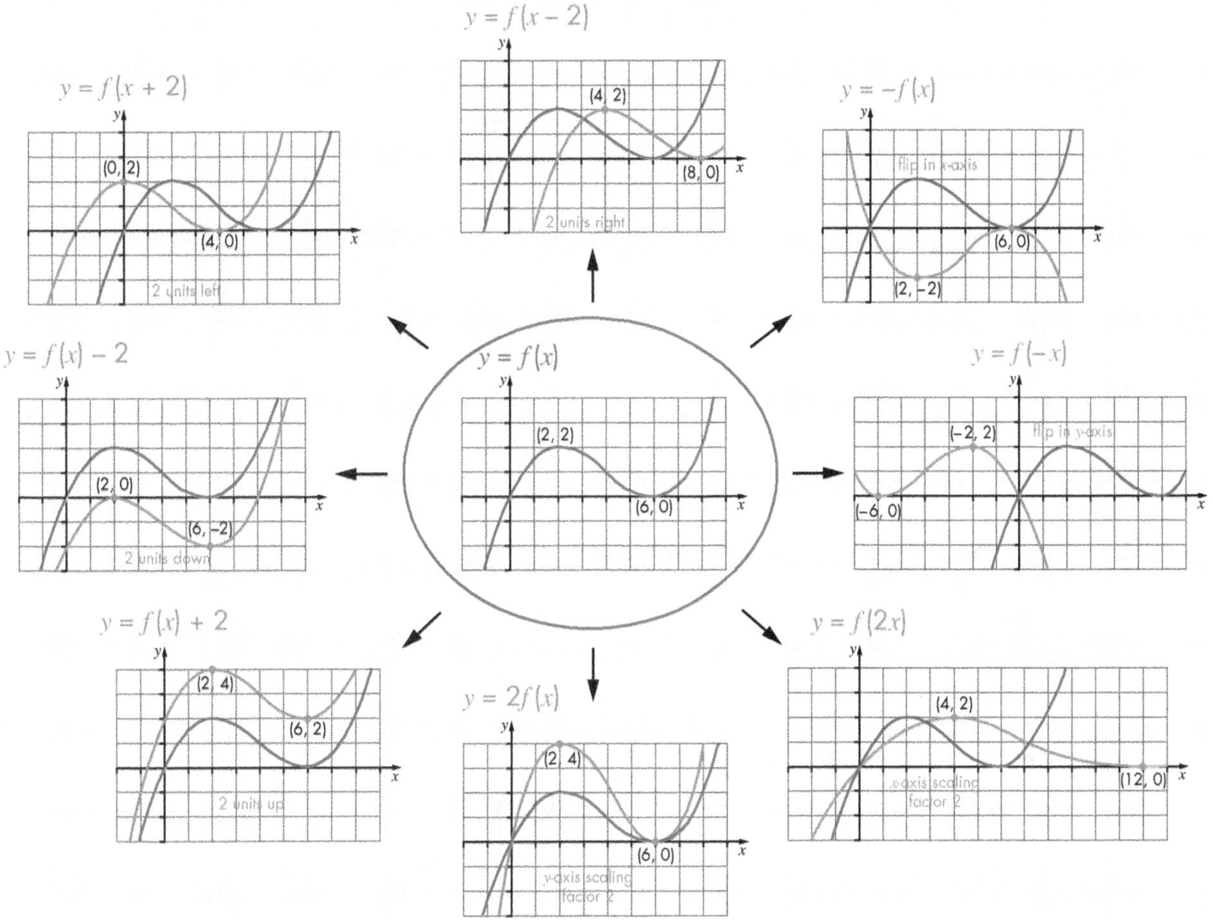

In Questions 1–6 the graph $y = f(x)$ is shown. Sketch the indicated graphs showing clearly the images of the named points.

1

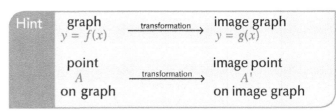

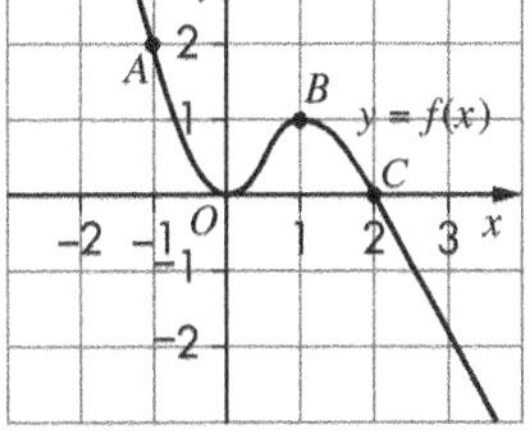

Hint

graph $y = f(x)$	transformation →	image graph $y = g(x)$
point A on graph	transformation →	image point A' on image graph

a $y = f(x + 2)$ **b** $y = f(x - 1)$

c $y = -f(x)$ **d** $y = f(x) - 1$

2

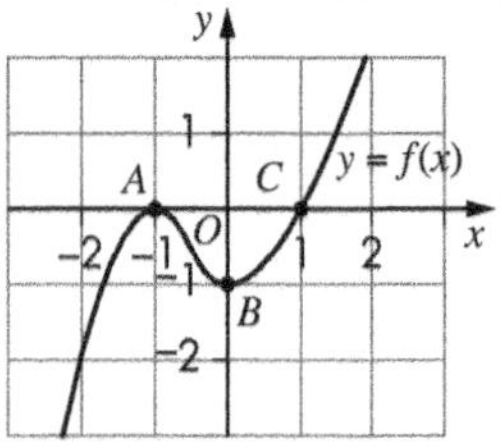

a $y = f(x - 2)$ **b** $y = f(x + 1)$

c $y = -f(x)$ **d** $y = f(x) + 1$

3

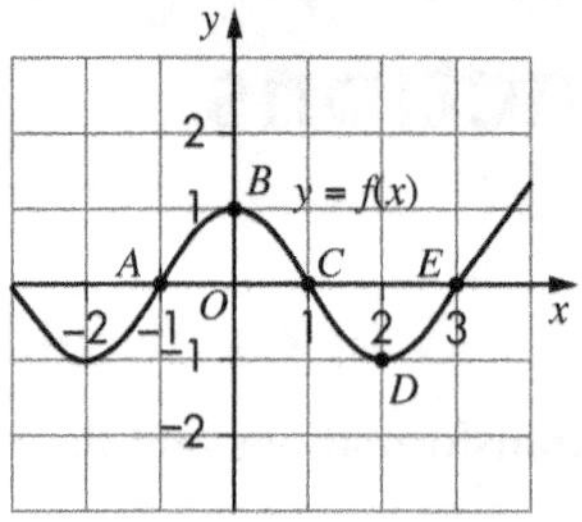

a $y = f(x + 3)$ **b** $y = f(x - 1)$

c $y = -f(x)$ **d** $y = f(x) - 1$

4

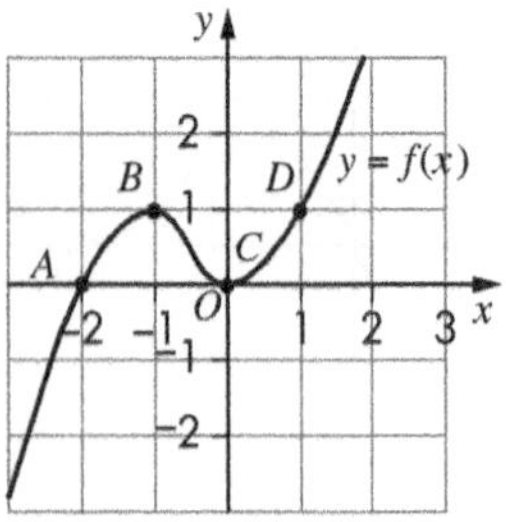

a $y = -f(x) + 1$ **b** $y = f(x + 1) - 1$

5

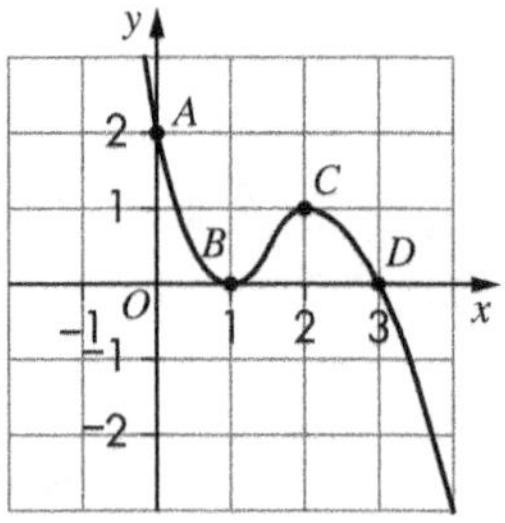

a $y = -f(x) + 2$ **b** $y = f(x + 1) - 1$

6

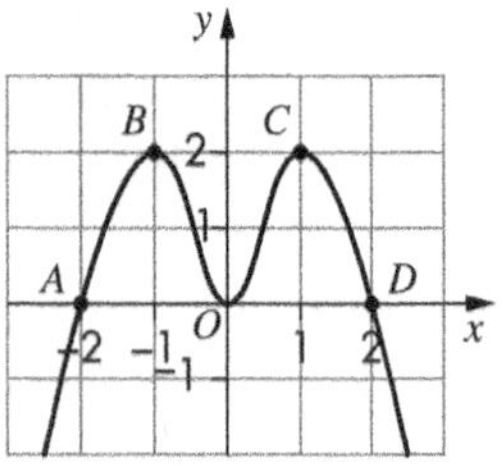

a $y = 1 - f(x)$

b $y = -f(x + 2)$

7 Find the value of a for each graph.

a

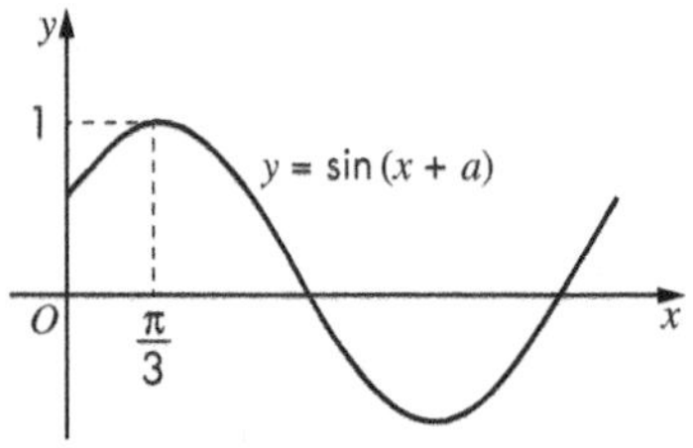

b

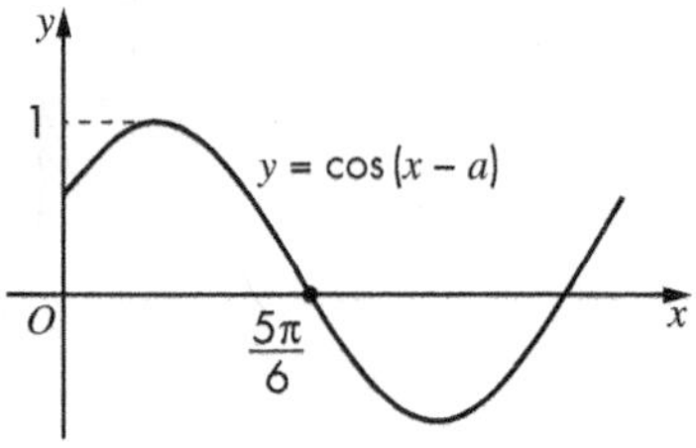

8 Find the values of a and b for each graph.

a

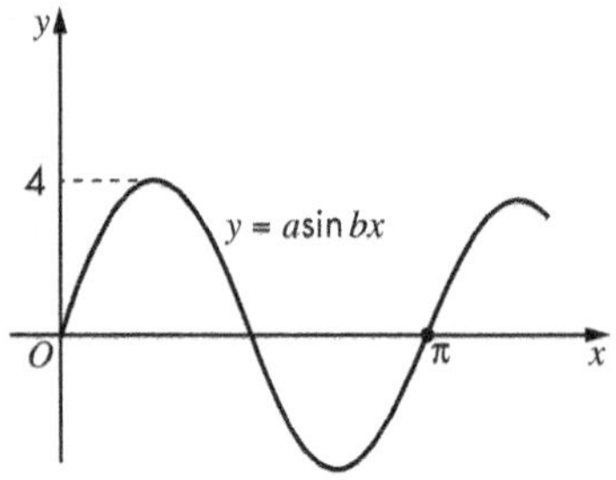

b

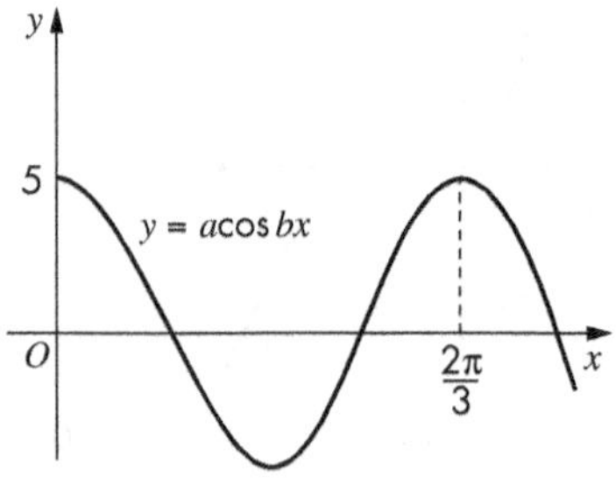

Exercise 3.2 Graphs of quadratic functions

Example

Write $3 + 8x - 2x^2$ in the form $a + b(x + c)^2$ where a, b and c are constants.

$3 + 8x - 2x^2 = 3 - 2x^2 + 8x$

$= 3 - 2(x^2 - 4x)$ — Rearrange to get an $(x^2 \pm x)$ term in brackets.

Consider $x^2 - 4x = (x - 2)(x - 2) - 4$ — Halve the x coefficient -4 to get -2. This leads to two equal factors, $x - 2$. $(x - 2)(x - 2) = x^2 - 4x + 4$ so remove the constant term $+4$ using -4.

$= (x - 2)^2 - 4$ — The square term is now complete.

So $3 - 2(x^2 - 4x) = 3 - 2[(x - 2)^2 - 4]$ — Replace $x^2 - 4x$ with $(x - 2)^2 - 4$

$= 3 - 2(x - 2)^2 + 8$ — Multiply out.

$= 11 - 2(x - 2)^2$ — Tidy up.

1 Express in the form $(x + b)^2 + c$.

a $x^2 + 4x + 3$ **b** $x^2 - 6x - 1$ **c** $x^2 + 3x + 1$

d $x^2 + 5x + 2$ **e** $x^2 - 5x + 2$ **f** $x^2 - x - 4$

g $x^2 + x - 7$ **h** $x^2 + 5x$ **i** $x^2 + 20x + 100$

Hint $(x + b)^2 + c$ is called the 'completed square form'.

2 Express $f(x)$ in the form $a(x^2 + bx) + c$.

a $f(x) = 2x^2 - 4x + 1$ **b** $f(x) = 3x^2 + 6x + 2$ **c** $f(x) = 5x^2 + 15x - 7$

d $f(x) = 2x^2 - 10x - 5$ **e** $f(x) = 4x^2 + 12x + 17$ **f** $f(x) = 7x^2 - 28x - 12$

g $f(x) = 2x^2 + 3x - 1$ **h** $f(x) = 3x^2 + 5x + 7$ **i** $f(x) = 2x^2 - x - 4$

3 Express $f(x)$ in the form $a(x + b)^2 + c$.

a $f(x) = 2x^2 + 4x + 1$ **b** $f(x) = 3x^2 - 6x - 2$

c $f(x) = (2x + 3)(2x + 5)$ **d** $f(x) = (2x - 1)(2x - 3)$

e $f(x) = (3x + 2)(3x + 4)$ **f** $f(x) = (5x - 1)(5x + 11)$

g $f(x) = (2x - 1)(x + 2)$ **h** $f(x) = 6(x + 1) + (5x - 1)(x - 2)$

4 Express in the form $p - q(x^2 + rx)$.

a $3 - 4x - 2x^2$ **b** $1 + 6x - 3x^2$ **c** $28x - 7x^2$ **d** $6x - x^2$

e $2 - 2x - x^2$ **f** $1 + 4x - x^2$ **g** $-5x^2 + 10x - 1$ **h** $-4x^2 - 16x$

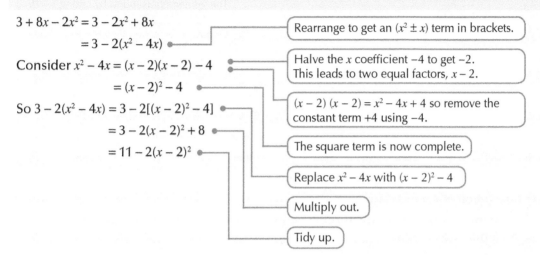

Example

Sketch the graph $y = (x - 3)^2 + 1$ showing the y-intercept and the coordinates of the minimum turning point.

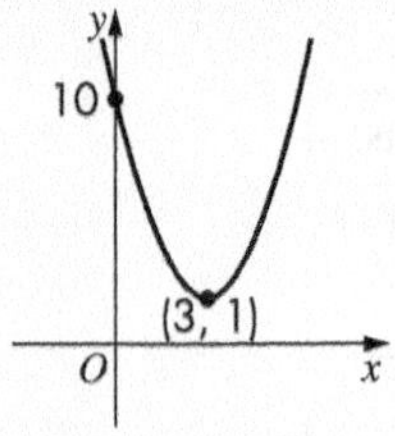

When $x = 0$, $y = (0 - 3)^2 + 1 = 9 + 1 = 10$ — The y-intercept is found by letting $x = 0$.

So the y-intercept is (0, 10).

$y = x^2 \rightarrow y = (x - 3)^2$ — Graph $y = x^2$ moves 3 units right.

$y = (x - 3)^2 \rightarrow y = (x - 3)^2 + 1$ — Graph $y = (x - 3)^2$ moves 1 unit up.

The minimum turning point is (3, 1).

Track the movement of (0,0):

(0, 0) ⟶ (3, 0) ⟶ (3, 1)

$y = x^2$ min ⟶ $y = (x - 3)^2 + 1$ min

5 For the graph of each equation $y = f(x)$:

i state the coordinates of the turning point and whether it is a maximum or minimum point

ii sketch the graph $y = f(x)$

iii annotate the graph to show the turning point and the y-axis intercept

iv state the maximum/minimum value for the function f.

a $y = (x + 1)^2 + 2$ **b** $y = (x - 3)^2 + 1$

c $y = 2 - (x - 1)^2$ **d** $y = 3 - (x + 2)^2$

e $y = 2(x - 1)^2 - 1$ **f** $y = 3(x + 2)^2 + 4$

g $y = -(x + 5)^2$ **h** $y = 2(x + 1)^2$

i $y = -1 - 5(x - 2)^2$ **j** $y = -5 - 2(x + 1)^2$

Hint If (a, b) is a max/min turning point of $y = f(x)$, then b is the max/min value of f.

Hint When the x^2 term is negative, the graph will be inverted (compared to the $y = x^2$ graph) and there will be a maximum turning point with the typical equation being $y = a - b(x + c)^2$.

6 Find the equation of each graph.

a

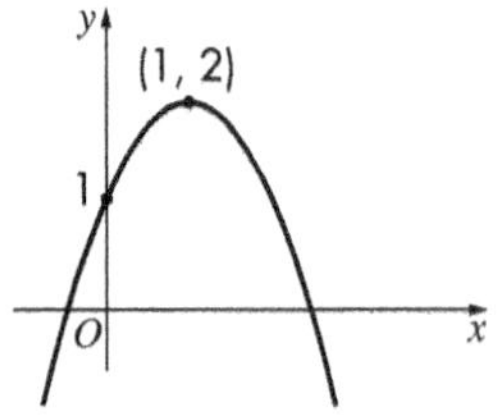

b

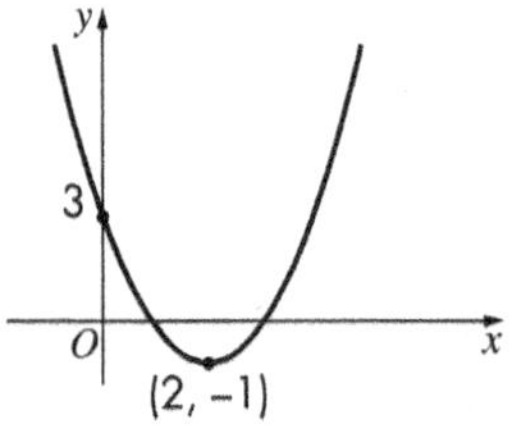

c

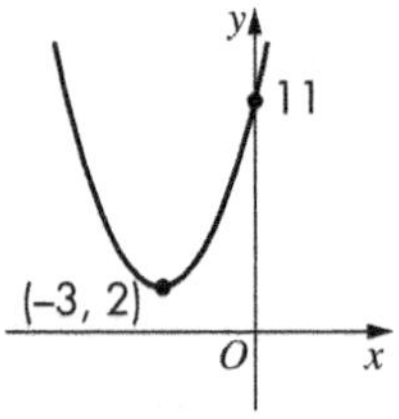

Exercise 3.3 Graphs of exponential and logarithmic functions

1 Sketch the graphs:

a $y = 2^x - 1$

b $y = 2^{x+1}$

c $y = 2^{x-1}$

d $y = -2^x$

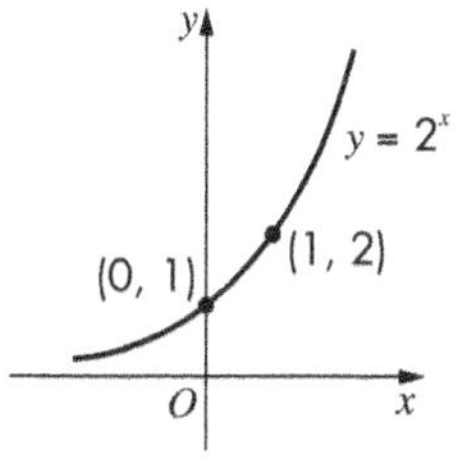

Hint Refer back to the diagrams at the top of page 9 where examples of transformations and their related graphs are shown.

2 Sketch the graphs:

a $y = \log_5 x + 1$

b $y = \log_5 (x + 1)$

c $y = -\log_5 x$

d $y = \log_5 (x - 2) - 1$

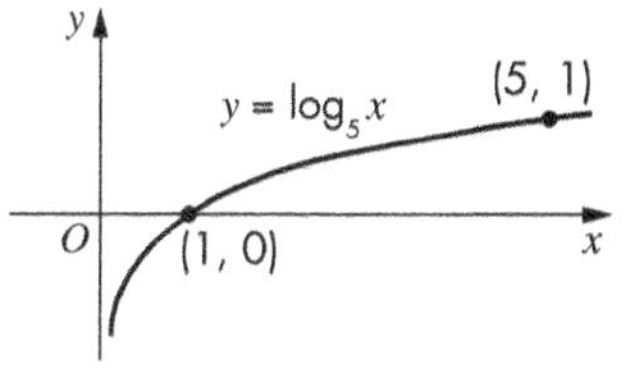

3 Each graph has equation $y = a^x$ where a is a constant. Find the equations.

a

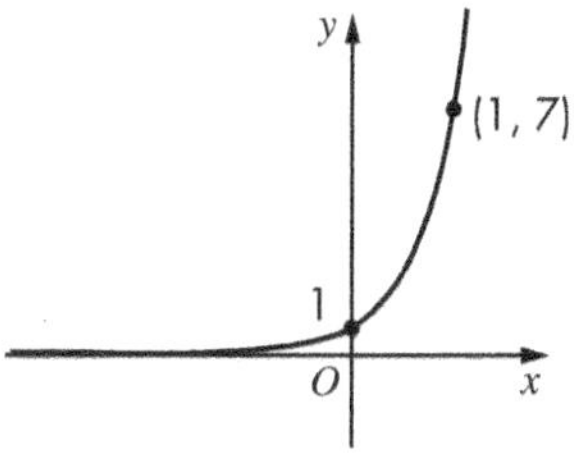

b

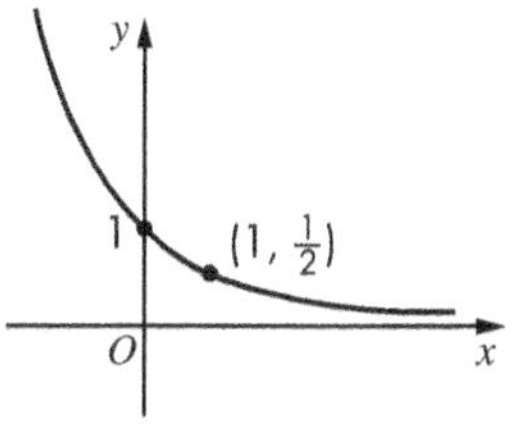

c

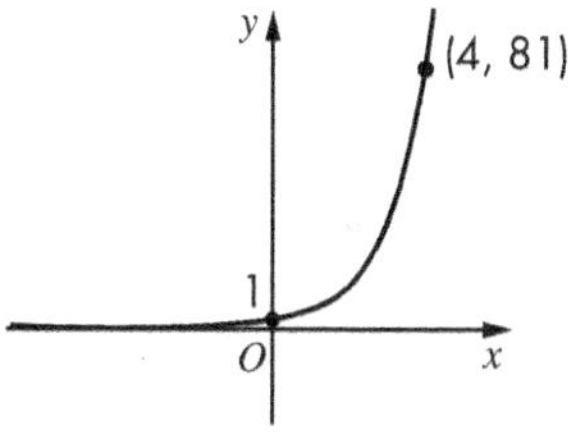

d

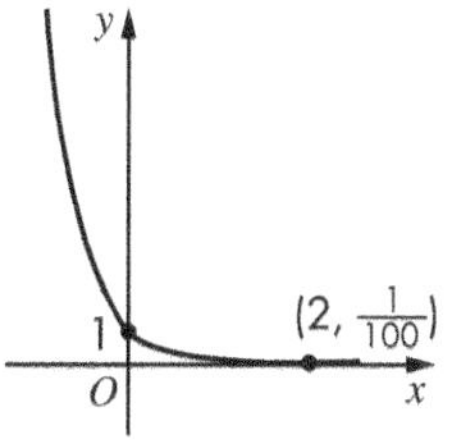

e

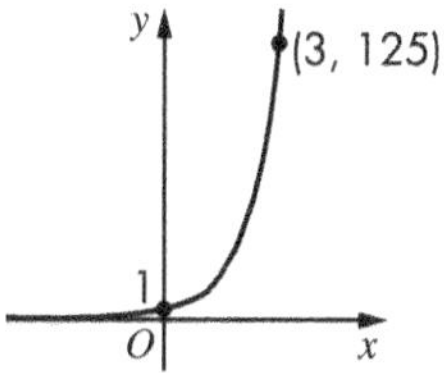

f

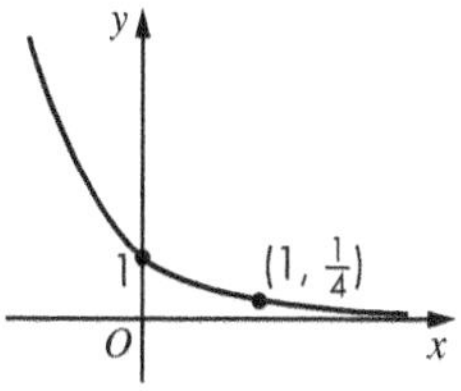

4 Each graph has equation $y = \log_a x$ where a is a constant. Find the equations.

a

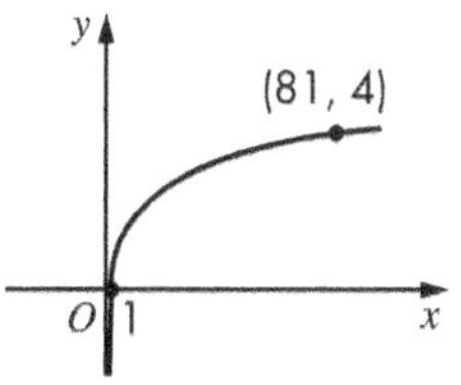

b

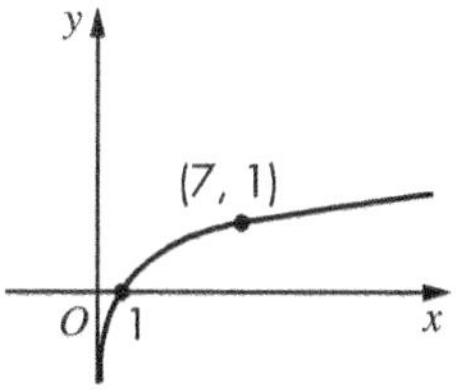

c

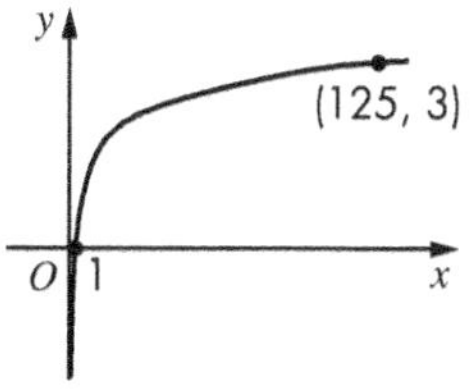

5 Sketch these graphs showing the axes intercepts and the coordinates of one other point:

a $y = \log_3 (x - 1)$ **b** $y = 3^{x-2}$ **c** $y = \log_2 x - 1$

d $y = 2^x + 1$ **e** $y = 3\log_4 x$ **f** $y = 2 \times 3^x$

Exercise 3.4 Graphs of trigonometric functions

1 Give the coordinates of the maximum and minimum stationary points on these graphs for $0 \leqslant x \leqslant 2\pi$:

a

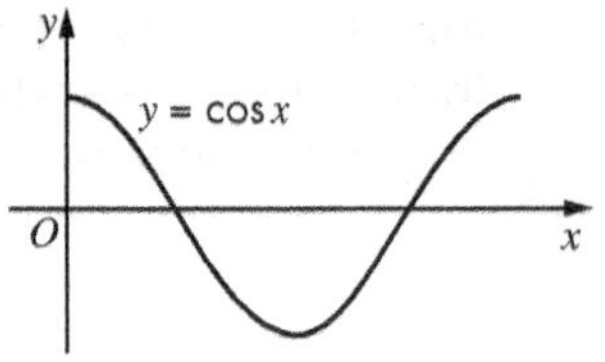

b

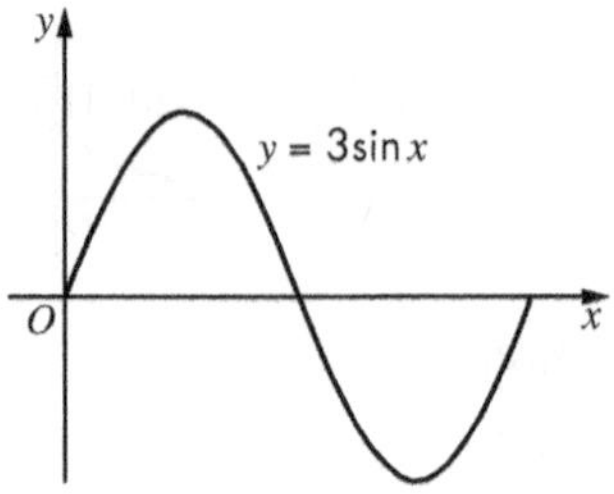

c $y = 2\cos x$

d $y = \frac{1}{2}\sin x$

e $y = 10\sin x$

f $y = \frac{1}{2}\cos x$

g

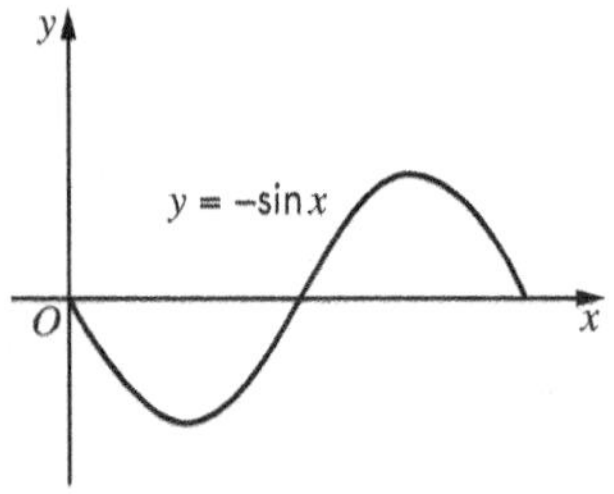

h

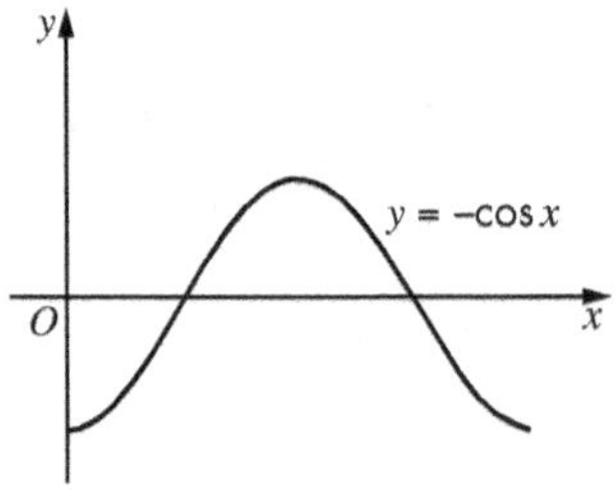

2 Find the coordinates of the maximum and minimum stationary points on each graph for $0 \leqslant x \leqslant 2\pi$.

a

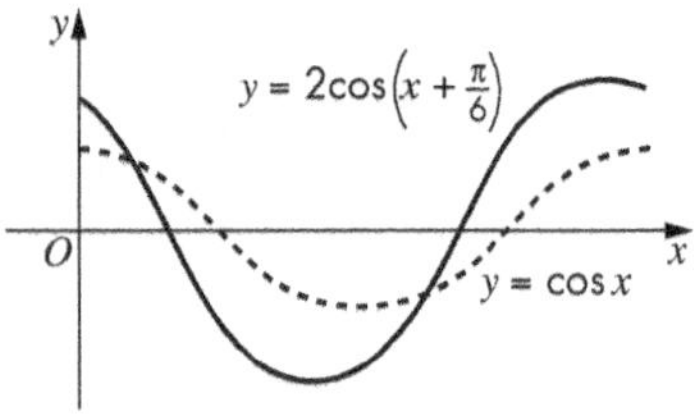

b

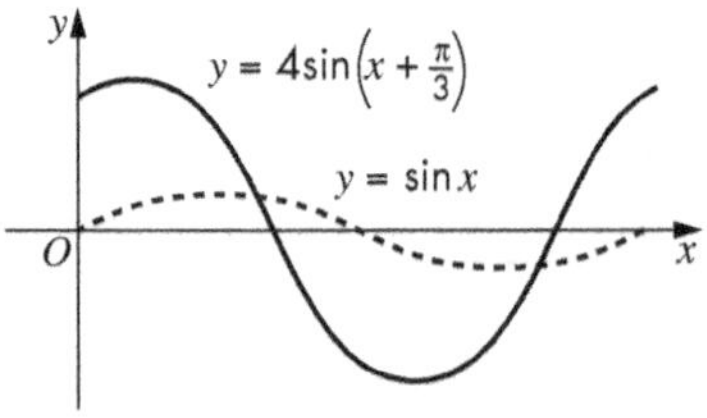

c

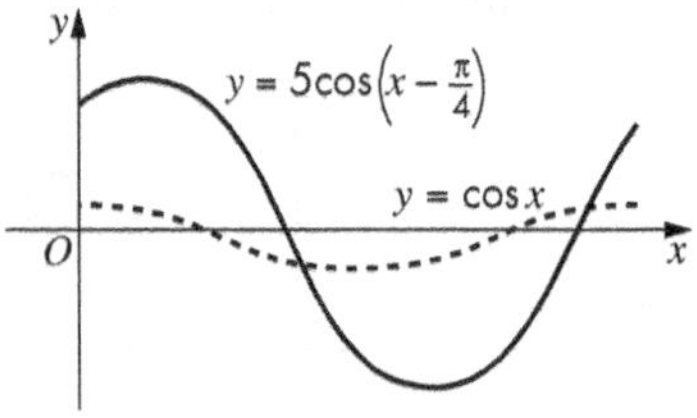

d

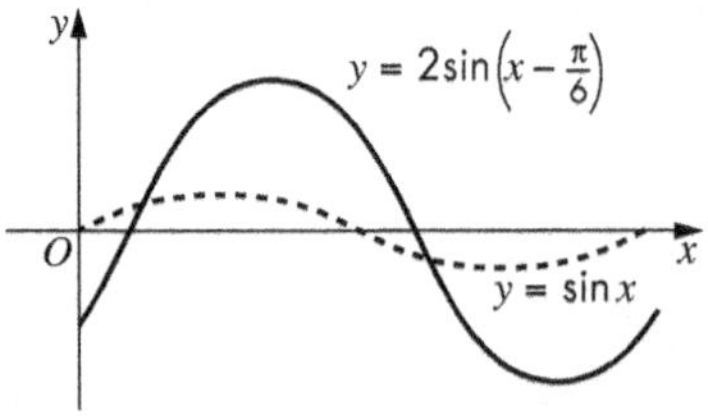

e $y = 15\sin\left(x - \frac{\pi}{4}\right)$

f $y = 6\cos\left(x + \frac{\pi}{3}\right)$

g $y = 3\sin\left(x + \frac{\pi}{6}\right)$

h $y = 10\cos\left(x - \frac{\pi}{5}\right)$

3 Sketch these graphs for $0 \leqslant x \leqslant 2\pi$:

a $y = 2\cos x$

b $y = \frac{1}{2}\sin 2x$

c $y = \cos\left(x - \frac{\pi}{6}\right)$

d $y = 2\sin\left(x + \frac{\pi}{3}\right)$

e $y = \sin 2x + 1$

f $y = \frac{1}{2}\cos 2x - 1$

Exercise 3.5 Graphs of the derived function

Reminder

Here is a cubic graph with its gradient graph:

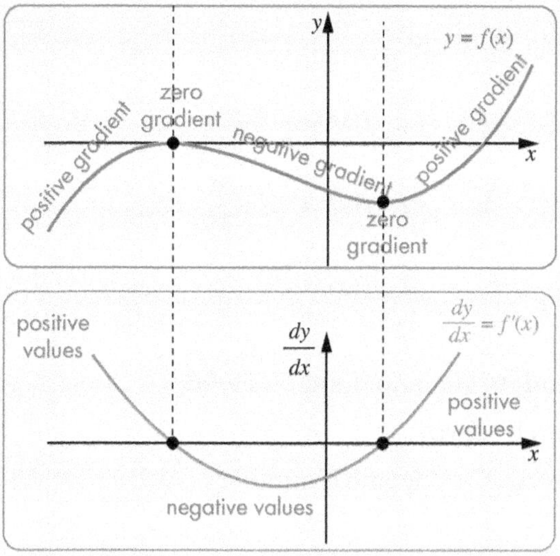

The graph $y = f(x)$ shows the values of the function f.

The graph $\frac{dy}{dx} = f'(x)$ shows the values of the gradient of the original function.

Differentiating a cubic (degree 3) will give a quadratic (degree 2) which explains the parabolic shape of the gradient graph.

1 For each of these quadratic graphs $y = f(x)$, sketch the graph of $y = f'(x)$:

a 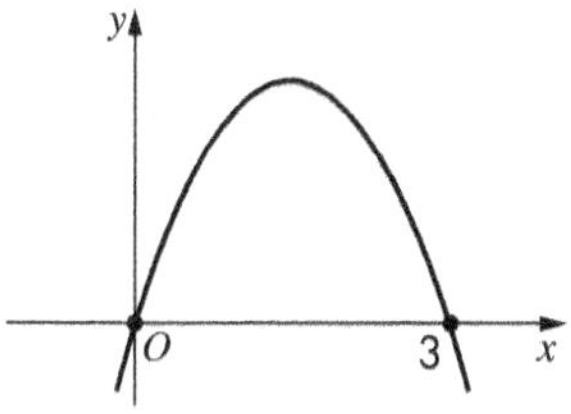

b 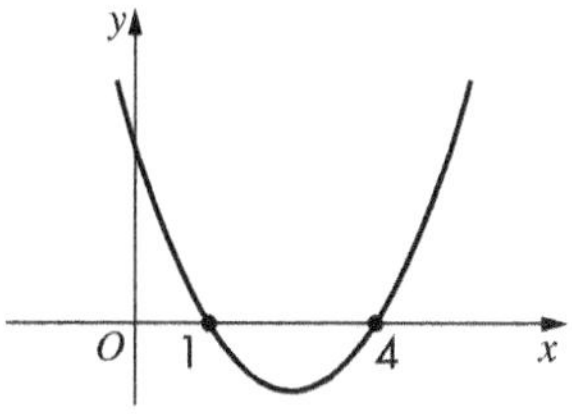

c 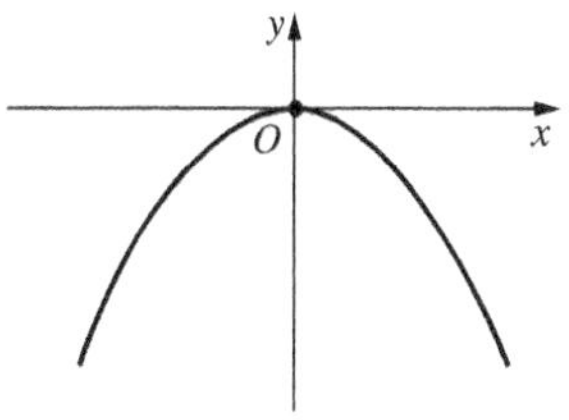

2 Sketch the graph $y = f'(x)$ for each of these cubic graphs $y = f(x)$:

a 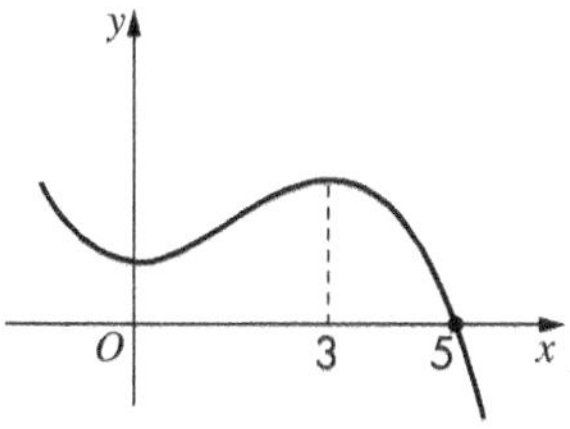

b 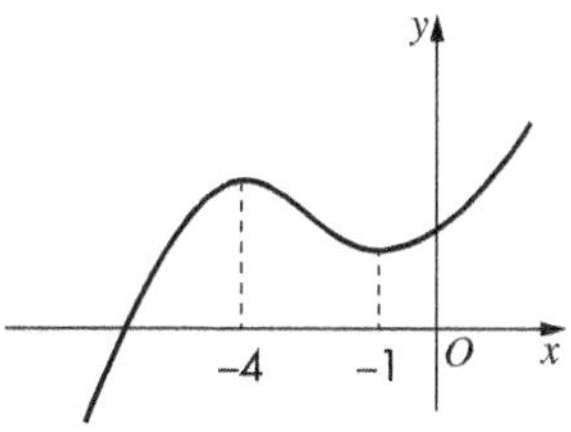

c 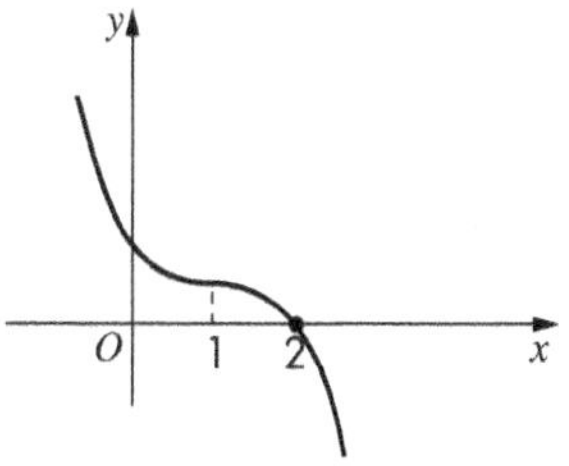

3 These are graphs of trigonometric functions. In each case sketch the derived graph.

a 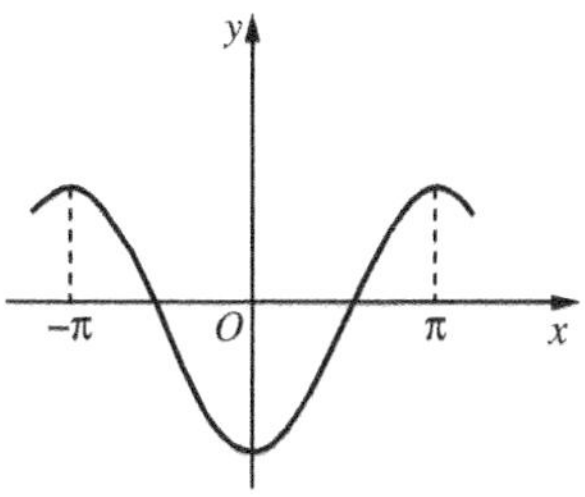

b 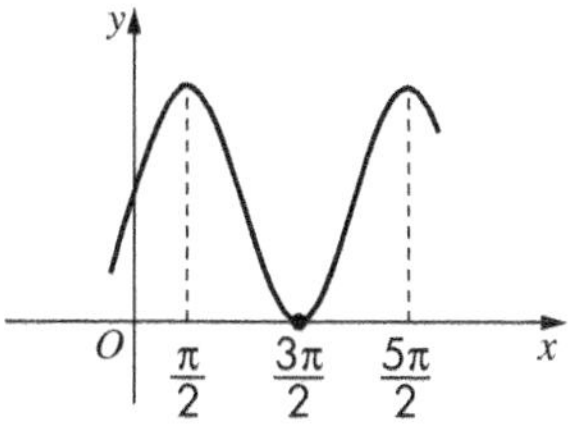

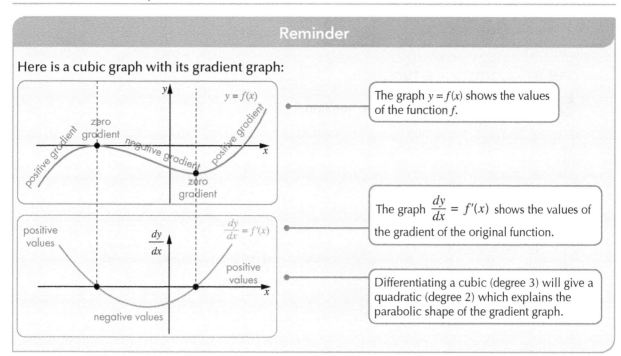

4 Determining composite and inverse functions

Exercise 4.1 Composite functions

1 State a suitable domain for the function f.

a $f(x) = \frac{1}{x}$ **b** $f(x) = \frac{x+1}{x+2}$ **c** $f(x) = \frac{x}{x-4}$

d $f(x) = \frac{3}{x^2-1}$ **e** $f(x) = \frac{x+2}{x^2-4}$ **f** $f(x) = \frac{3x+2}{x^2-16}$

g $f(x) = \frac{2x-1}{3x^2}$ **h** $f(x) = \frac{x-5}{9-4x^2}$ **i** $f(x) = \sqrt{x-4}$

j $f(x) = \sqrt{x+3}$

k $f(x) = \sqrt{10-x}$

l $f(x) = \sqrt{5-2x}$

Hint When finding suitable domains **avoid**:
- division by zero
- square roots of negative numbers.

Example

If $f(x) = 2x - 3$ and $g(x) = 2x^2$, find $f(g(x))$ and $g(f(x))$ and show $2f(g(x)) - g(f(x)) = 24(x-1)$

$f(g(x)) = f(2x^2)$ — Replace $g(x)$ with $2x^2$. ('Inside function' first.)

$= 2(2x^2) - 3 = 4x^2 - 3$ — $f(x) = 2x - 3$ so replace x with $2x^2$ and then simplify.

$g(f(x)) = g(2x-3)$ — Replace $f(x)$ with $2x - 3$.

$= 2(2x-3)^2 = 8x^2 - 24x + 18$ — $g(x) = 2x^2$ so replace x with $2x - 3$ and then simplify.

So $2f(g(x)) - g(f(x))$

$= 2(4x^2 - 3) - (8x^2 - 24x + 18)$ — Replace $f(g(x))$ and $g(f(x))$ with $4x^2 - 3$ and $8x^2 - 24x + 18$.

$= 8x^2 - 6 - 8x^2 + 24x - 18$

$= 24x - 24 = 24(x-1)$ as required. — Simplify and rearrange.

2 $f(x) = 4x - 1$. Find:

a $f(x^2)$ **b** $f(2x)$ **c** $f(x+1)$ **d** $f(3-x)$

3 $g(x) = x^2 + 2$. Find:

a $g(2x)$ **b** $g(x-1)$ **c** $g(3x+1)$ **d** $g(2-5x)$

4 $f(x) = 3x - 1$ and $g(x) = x^2$. Find:

a $g(f(x))$ **b** $f(g(x))$ **c** $f(f(x))$ **d** $g(g(x))$

5 Find **i** $g(f(x))$ and **ii** $f(g(x))$ when:

a $f(x) = 3x$ and $g(x) = 2x + 1$

b $f(x) = x^2$ and $g(x) = 2x$

c $f(x) = x - 5$ and $g(x) = x^2$

d $f(x) = 1 + 2x$ and $g(x) = x^2 - 2$

6 $f(x) = 2 - x$ and $g(x) = 2x + k$.

a Find $f(g(x))$.

b Find $g(f(x))$.

c Show that $f(g(x)) + g(f(x)) = 6 - 4x$

d Find $g(f(x)) - f(g(x))$.

7 $f(x) = 2x$ and $g(x) = x^2 - m$.

a Show that $g(f(x)) - f(g(x)) = 2x^2 + m$

b Show that $(f(x))^2 - g(f(x)) = m$

8 Simplify by writing as a single fraction:

a $2 \times \frac{1}{x} - 1$

b $1 - 4 \times \frac{1}{x}$

c $2 \times \frac{1}{x-1} + 1$

d $2 \times \frac{1}{x-1} - 1$

e $1 + 4 \times \frac{1}{x+1}$

f $1 - 3 \times \frac{1}{x-1}$

g $4 \times \frac{1}{x^2-9} + 1$

h $3 \times \frac{1}{x^2-1} - 1$

Hint $1 - 5 \times \frac{1}{x+2} = \frac{x+2}{x+2} - \frac{5}{x+2} = \frac{x+2-5}{x+2} = \frac{x-3}{x+2}$

9 For the functions f and g, defined on suitable domains, find an expression for $h(x)$ where $h(x) = g(f(x))$. Give your answer as a single fraction.

a $f(x) = \frac{1}{x}$ and $g(x) = 2x - 1$

b $f(x) = \frac{1}{x+1}$ and $g(x) = 3x + 1$

c $f(x) = \frac{1}{x-1}$ and $g(x) = 1 - 2x$

d $f(x) = \frac{1}{x^2-1}$ and $g(x) = 2x + 1$

e $f(x) = \frac{1}{x^2-9}$ and $g(x) = 3x - 1$

f $f(x) = \frac{1}{x^2-1}$ and $g(x) = 1 + 2x$

Exercise 4.2 Inverse functions and graphs

1 Find a formula for the inverse of these functions:

a $f(x) = 2x + 4$

b $g(x) = 3 - x$

c $f(x) = 6 - 2x$

d $h(x) = x - \frac{1}{2}$

e $f(x) = 2(x + 1)$

f $g(x) = 4\left(1 - \frac{1}{2}x\right)$

g $f(x) = \frac{1}{x}$

h $h(x) = \frac{1}{x+1}$

i $f(x) = \frac{3}{x-1}$

j $f(x) = \frac{1}{x} + 1$

k $g(x) = 2 - \frac{1}{x}$

l $f(x) = 1 + \frac{1}{x+1}$

Hint Here are the graphs of three functions and their inverses:

Doubling

$f(x) = 2x$

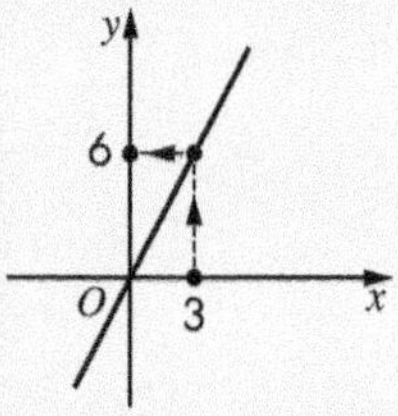

⇓

Halving

$f^{-1}(x) = \frac{1}{2}x$

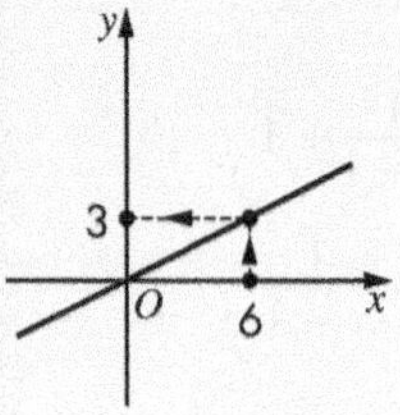

Squaring

$f(x) = x^2\ (x \geqslant 0)$

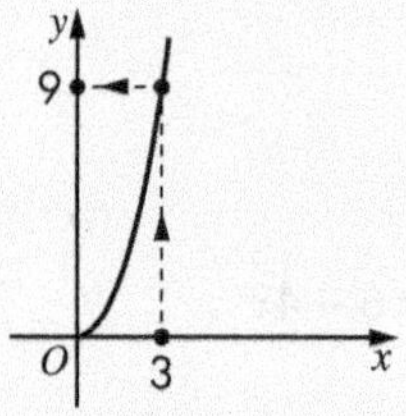

⇓

Square-rooting

$f^{-1}(x) = \sqrt{x}\ (x \geqslant 0)$

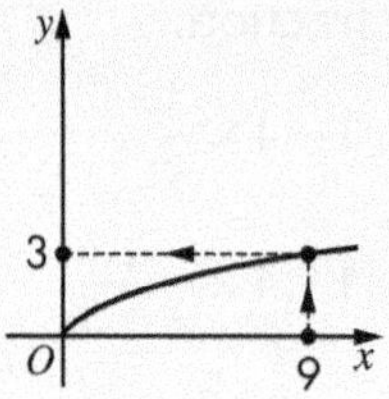

Inverting

$f(x) = \dfrac{1}{x}\ (x \neq 0)$

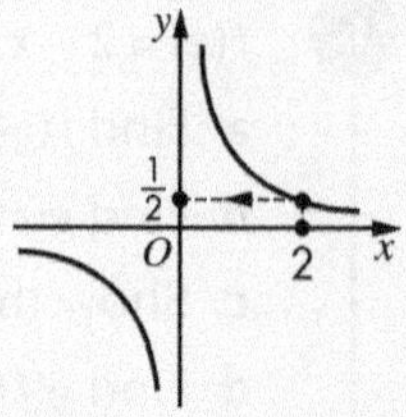

⇓

Inverting

$f^{-1}(x) = \dfrac{1}{x}\ (x \neq 0)$

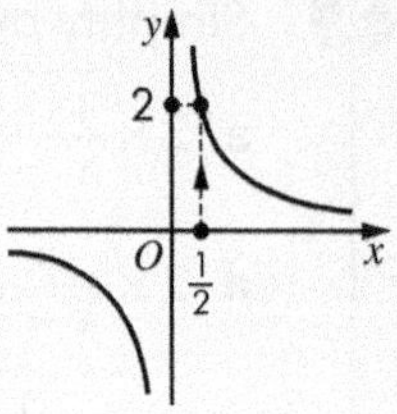

f is its own inverse.

Flip the graph $y = f(x)$ in the line $y = x$ to get the graph $y = f^{-1}(x)$.

2 Match each $y = f(x)$ graph with the correct $y = f^{-1}(x)$ inverse graph.

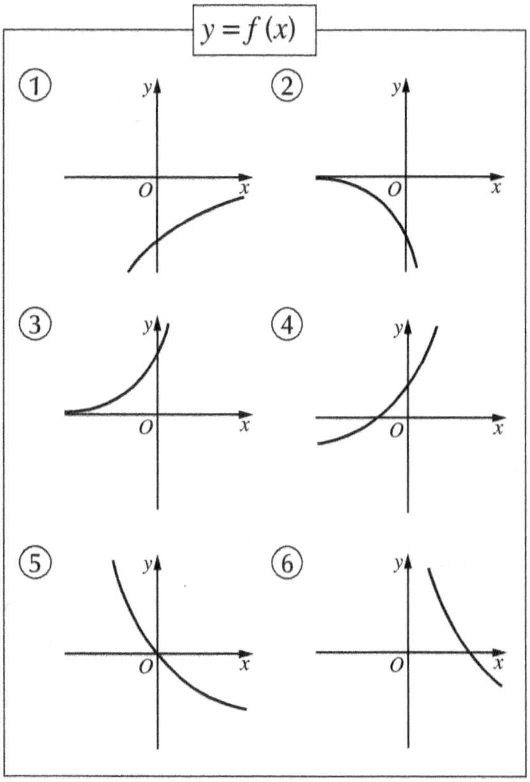

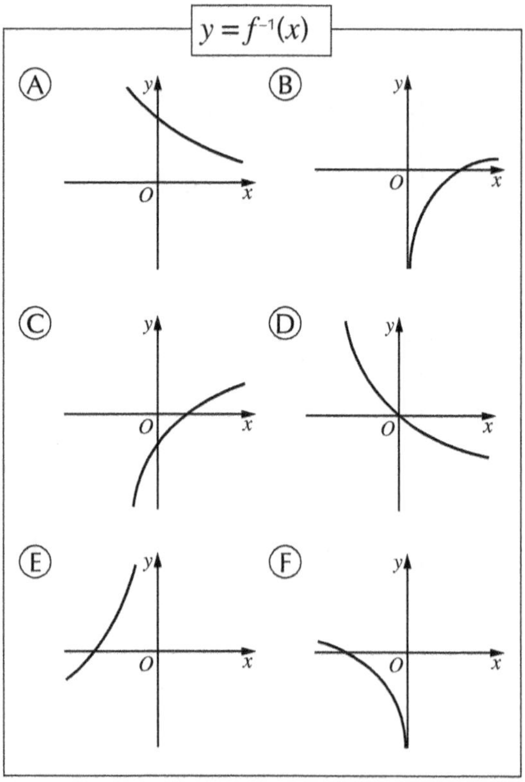

5 Determining vector connections

Exercise 5.1 Coordinates, components and unit vectors

1. Write each of these in column form:

a $3\mathbf{i} - 2\mathbf{j} + 5\mathbf{k}$ b $4\mathbf{i} - \sqrt{2}\mathbf{j}$ c $-\mathbf{i} - 2\mathbf{j} + 8\mathbf{k}$ d $-\sqrt{5}\mathbf{j} - 2\mathbf{k}$

e $\mathbf{i} + \mathbf{j} + \mathbf{k}$ f $\mathbf{i} - \mathbf{j} - \mathbf{k}$ g $18\mathbf{j} - \sqrt{11}\mathbf{k}$ h $-\sqrt{2}\mathbf{j}$

2. For the points $A(-1, 2, 3)$, $B(6, 0, 1)$, $C(4, -1, 8)$ and $D(0, 2, -2)$, find the following in component form.

a $\overrightarrow{OA}$ b $\overrightarrow{OB}$ c $\overrightarrow{OC}$ d $\overrightarrow{OD}$

3. For the points $P(3, -1, 2)$, $Q(3, 0, -1)$, $R(0, -4, 5)$ and $S(4, -4, 0)$ find the following in terms of **i**, **j** and **k**.

a $\overrightarrow{OP}$ b $\overrightarrow{OQ}$ c $\overrightarrow{OR}$ d $\overrightarrow{OS}$

4. For these square-based stepped pyramids, find the coordinates of the marked points:

a

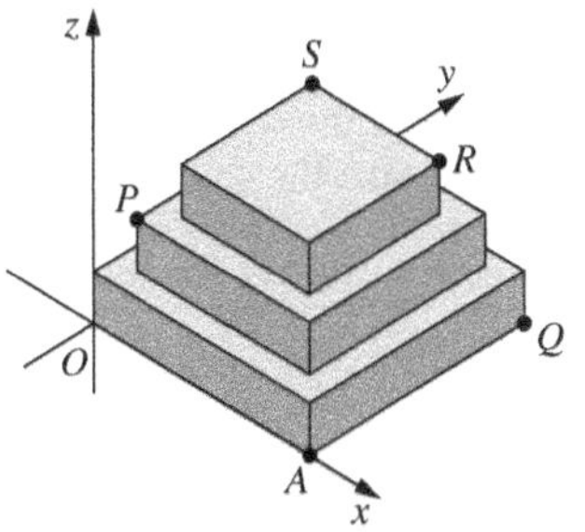

The step height is 2 units.

The step width is 1 unit.

A is the point (10, 0, 0).

b

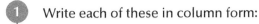

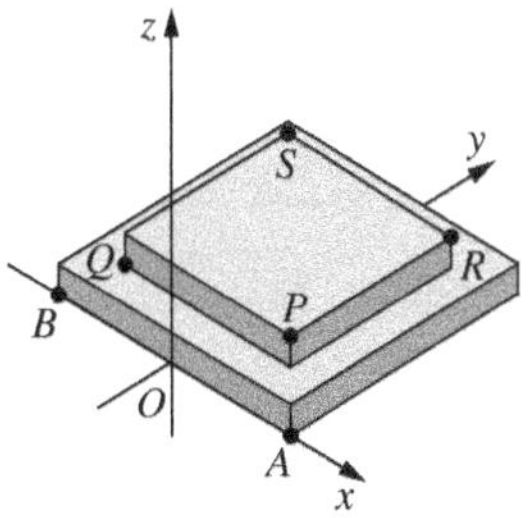

The step height is 3 units.

The step width is 4 units.

A is the point (15, 0, 0).

B is the point (−15, 0, 0).

5. Use the diagram from Question 4 part **a** to find the following in component form.

a $\overrightarrow{OQ}$ b $\overrightarrow{OS}$ c $\overrightarrow{AQ}$

d $\overrightarrow{QS}$ e $\overrightarrow{RP}$ f $\overrightarrow{SA}$

6. Use the diagram from Question 4 part **b** to find the following in component form.

a $\overrightarrow{QB}$ b $\overrightarrow{PO}$ c $\overrightarrow{RS}$

d $\overrightarrow{AS}$ e $\overrightarrow{BR}$ f $\overrightarrow{SQ}$

7. Rewrite the vectors from Question 5 in terms of **i**, **j** and **k**.

8. Rewrite the vectors from Question 6 in terms of **i**, **j** and **k**.

Exercise 5.2 Working with vectors and vector magnitude

Example

Calculate the magnitude of $\mathbf{a} - 2\mathbf{b}$ where $\mathbf{a} = \begin{pmatrix} -1 \\ 2 \\ 2 \end{pmatrix}$ and $\mathbf{b} = \begin{pmatrix} 3 \\ -2 \\ 1 \end{pmatrix}$

$\mathbf{a} - 2\mathbf{b} = \begin{pmatrix} -1 \\ 2 \\ 2 \end{pmatrix} - 2\begin{pmatrix} 3 \\ -2 \\ 1 \end{pmatrix}$ — Replace **a** and **b** with their component form.

$= \begin{pmatrix} -1 \\ 2 \\ 2 \end{pmatrix} - \begin{pmatrix} 6 \\ -4 \\ 2 \end{pmatrix}$ — Multiply the components of **b** by the scalar 2.

$= \begin{pmatrix} -7 \\ 6 \\ 0 \end{pmatrix}$ — Subtract the corresponding components.

So $|\mathbf{a} - 2\mathbf{b}| = \left|\begin{pmatrix} -7 \\ 6 \\ 0 \end{pmatrix}\right|$ — The lines | | mean 'the magnitude of'.

$= \sqrt{(-7)^2 + 6^2 + 0^2}$ — This is the sum of the squares of the components. Don't forget to take the square root.

$= \sqrt{49 + 36 + 0} = \sqrt{85}$ — This is the exact value of the magnitude.

1 Calculate the magnitude (length) of these vectors:

a $\begin{pmatrix} 1 \\ -2 \\ 2 \end{pmatrix}$ **b** $\begin{pmatrix} -4 \\ 8 \\ -1 \end{pmatrix}$ **c** $\begin{pmatrix} 5 \\ -10 \\ -10 \end{pmatrix}$ **d** $\begin{pmatrix} -3 \\ 0 \\ 4 \end{pmatrix}$ **e** $\begin{pmatrix} \sqrt{2} \\ -1 \\ \sqrt{2} \end{pmatrix}$

f $\begin{pmatrix} 0 \\ -5 \\ 0 \end{pmatrix}$ **g** $\begin{pmatrix} 1 \\ -2 \\ -3 \end{pmatrix}$ **h** $\begin{pmatrix} \sqrt{3} \\ -1 \\ -1 \end{pmatrix}$ **i** $\begin{pmatrix} \frac{1}{4} \\ -1 \\ 2 \end{pmatrix}$

2 Calculate the magnitude of these vectors:

a $2\mathbf{i} - 2\mathbf{j} + \mathbf{k}$ **b** $-3\mathbf{i} + 6\mathbf{j} - 2\mathbf{k}$ **c** $4\mathbf{i} - 4\mathbf{j} + 7\mathbf{k}$ **d** $4\mathbf{i} - 3\mathbf{k}$

e $\mathbf{i} - 5\mathbf{j} + \sqrt{10}\mathbf{k}$ **f** $\sqrt{5}\mathbf{i} + 4\mathbf{j} - 2\mathbf{k}$ **g** $2\mathbf{j} - \sqrt{5}\mathbf{k}$ **h** $2\sqrt{2}\mathbf{i} - \mathbf{k}$

i $\sqrt{5}\mathbf{j}$ **j** $\sqrt{3}\mathbf{i} - \mathbf{j} - 2\sqrt{3}\mathbf{k}$

3 $\mathbf{p} = \begin{pmatrix} -1 \\ 2 \\ 0 \end{pmatrix}$, $\mathbf{q} = \begin{pmatrix} 3 \\ 4 \\ -1 \end{pmatrix}$, $\mathbf{r} = \begin{pmatrix} -1 \\ -2 \\ 4 \end{pmatrix}$ and $\mathbf{s} = \begin{pmatrix} -2 \\ 0 \\ 2 \end{pmatrix}$

Calculate the following, giving your answer in component form and also in terms of **i, j** and **k**.

a $2\mathbf{p} - \mathbf{q}$ **b** $\mathbf{r} - 2\mathbf{s}$ **c** $\frac{1}{2}\mathbf{s} + \mathbf{p}$ **d** $3\mathbf{r} - 2\mathbf{p}$

e $-\mathbf{q} + \mathbf{p}$ **f** $-\mathbf{p} - \mathbf{r}$ **g** $2(\mathbf{r} - \mathbf{p})$ **h** $\mathbf{p} + \mathbf{r} + \mathbf{s}$

4 Using the vectors from Question 3 calculate the exact value of:

a $|\mathbf{q}|$ **b** $|\mathbf{r} - \mathbf{s}|$ **c** $|\mathbf{p} + \mathbf{q} + \mathbf{r}|$ **d** $|\mathbf{s} - 2\mathbf{p} + \mathbf{q}|$

5 Find a unit vector parallel to $\mathbf{u} + \mathbf{v}$ where:

a $\mathbf{u} = 3\mathbf{i} - \mathbf{j} + 2\mathbf{k}$ and $\mathbf{v} = -4\mathbf{i} - \mathbf{j}$

b $\mathbf{u} = \mathbf{i} - 5\mathbf{j} - 2\mathbf{k}$ and $\mathbf{v} = \mathbf{i} + \mathbf{j} + 6\mathbf{k}$

c $\mathbf{u} = -2\mathbf{i} + 5\mathbf{j} - 3\mathbf{k}$ and $\mathbf{v} = -4\mathbf{i} - 2\mathbf{j} + \mathbf{k}$

Hint To find a unit vector parallel to $\mathbf{p}$ multiply its components by $\frac{1}{|\mathbf{p}|}$.

Exercise 5.3 Working with position vectors

Reminder

$\overrightarrow{OP}$, written $\mathbf{p}$, is the **position vector** of the point P.

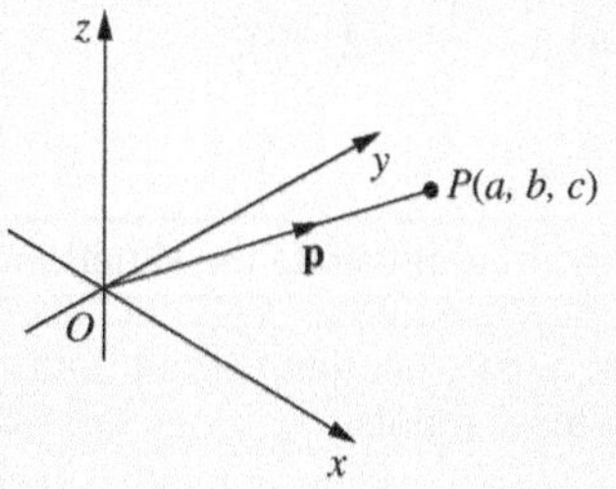

$P(a, b, c)$ ⟷ $\mathbf{p} = \begin{pmatrix} a \\ b \\ c \end{pmatrix}$

a, b and c are the coordinates of the point P.

a, b and c are the components of the position vector $\mathbf{p}$.

Think of $P(a, b, c)$ as the address of point P and $\mathbf{p} = \begin{pmatrix} a \\ b \\ c \end{pmatrix}$ as the instructions to get from the origin O to point P.

1 Give the position vectors of these points:

a $A(1, 2, 3)$ b $B(-1, 3, 4)$ c $C\left(0, -1, \frac{1}{2}\right)$ d $D(-3, 0, 0)$ e $E\left(\frac{1}{2}, -\frac{1}{2}, \frac{3}{2}\right)$

2 Give the coordinates of the points for which these are the position vectors:

a $\mathbf{t} = \begin{pmatrix} -1 \\ 2 \\ 3 \end{pmatrix}$ b $\mathbf{u} = \begin{pmatrix} 5 \\ 0 \\ -1 \end{pmatrix}$ c $\mathbf{a} = \begin{pmatrix} \frac{2}{3} \\ 0 \\ -\frac{1}{3} \end{pmatrix}$ d $\mathbf{b} = \begin{pmatrix} \frac{1}{2} \\ -\frac{1}{2} \\ 0 \end{pmatrix}$ e $\mathbf{q} = \begin{pmatrix} -5 \\ \frac{1}{2} \\ 4 \end{pmatrix}$

3 Write each 'journey' in terms of position vectors.

a $\overrightarrow{BC}$ b $\overrightarrow{LM}$ c $\overrightarrow{DE}$ d $\overrightarrow{AR}$ e $\overrightarrow{PT}$

f $\overrightarrow{OA}$ (careful!) g $\overrightarrow{KL}$ h $\overrightarrow{VW}$ i $\overrightarrow{CO}$

4 Find the components of:

a $\overrightarrow{AB}$ where A is $(-1, 0, 4)$ and B is $(6, -1, 3)$

b $\overrightarrow{RL}$ where R is $(2, 0, 0)$ and L is $(-1, -2, 3)$

c $\overrightarrow{MN}$ where M is $(2, 5, -1)$ and N is $(-2, -5, 1)$

d $\overrightarrow{AB}$ where A is $\left(0, \frac{1}{2}, \frac{3}{2}\right)$ and B is $\left(-1, -\frac{1}{2}, \frac{1}{2}\right)$

e $\overrightarrow{OC}$ where C is $(-1, 2, 6)$ (O is the origin)

f $\overrightarrow{TS}$ where T is $(-1, -7, 11)$ and S is $(4, 3, 3)$

g $\overrightarrow{EF}$ where E is $\left(\frac{1}{2}, 2, -1\right)$ and F is $\left(1, \frac{3}{2}, -\frac{1}{2}\right)$

h $\overrightarrow{KT}$ where K is $\left(\sqrt{2}, 0, \sqrt{2}\right)$ and T is $\left(2\sqrt{2}, -\sqrt{2}, \sqrt{2}\right)$

5 A is $(-1, 2, 1)$, B is $(2, 8, 4)$ and C is $(0, -3, 3)$. Use position vectors to find the components of:

a $\overrightarrow{AB}$ **b** $\overrightarrow{AC}$ **c** $\overrightarrow{BC}$ **d** $\overrightarrow{CA}$

6 The points $P(-1, 0, 2)$, $Q(3, 5, -2)$, $R(-1, 6, -7)$ and $S(-5, 1, -3)$ form a quadilateral.

a Find the components of:

i $\overrightarrow{PQ}$ **ii** $\overrightarrow{SR}$ **iii** $\overrightarrow{QR}$ **iv** $\overrightarrow{PS}$

b What sort of quadilateral is $PQRS$?

Example

Q divides PR in the ratio 3:2 where P is $(-1, 3, 5)$ and R is $(4, -2, 15)$.

Find the coordinates of Q.

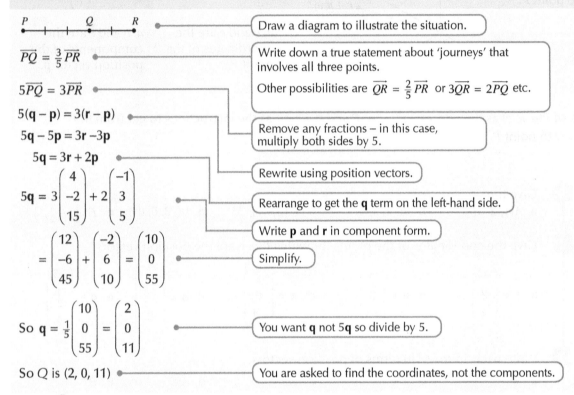

Draw a diagram to illustrate the situation.

$\overrightarrow{PQ} = \frac{3}{5}\overrightarrow{PR}$ — Write down a true statement about 'journeys' that involves all three points.

Other possibilities are $\overrightarrow{QR} = \frac{2}{5}\overrightarrow{PR}$ or $3\overrightarrow{QR} = 2\overrightarrow{PQ}$ etc.

$5\overrightarrow{PQ} = 3\overrightarrow{PR}$ — Remove any fractions – in this case, multiply both sides by 5.

$5(\mathbf{q} - \mathbf{p}) = 3(\mathbf{r} - \mathbf{p})$ — Rewrite using position vectors.

$5\mathbf{q} - 5\mathbf{p} = 3\mathbf{r} - 3\mathbf{p}$

$5\mathbf{q} = 3\mathbf{r} + 2\mathbf{p}$ — Rearrange to get the **q** term on the left-hand side.

$$5\mathbf{q} = 3\begin{pmatrix} 4 \\ -2 \\ 15 \end{pmatrix} + 2\begin{pmatrix} -1 \\ 3 \\ 5 \end{pmatrix}$$

Write **p** and **r** in component form.

$$= \begin{pmatrix} 12 \\ -6 \\ 45 \end{pmatrix} + \begin{pmatrix} -2 \\ 6 \\ 10 \end{pmatrix} = \begin{pmatrix} 10 \\ 0 \\ 55 \end{pmatrix}$$

Simplify.

$$\text{So } \mathbf{q} = \frac{1}{5}\begin{pmatrix} 10 \\ 0 \\ 55 \end{pmatrix} = \begin{pmatrix} 2 \\ 0 \\ 11 \end{pmatrix}$$

You want **q** not 5**q** so divide by 5.

So Q is $(2, 0, 11)$ — You are asked to find the coordinates, not the components.

7 Find the coordinates of:

a P which divides AB in the ratio 1:2 where A is $(-1, 0, 4)$ and B is $(5, 3, -2)$

b Q which divides CD in the ratio 2:1 where C is $(1, -3, 4)$ and D is $(7, 3, 1)$

c R which divides EF in the ratio 2:3 where E is $(-5, -2, 1)$ and F is $(0, 8, -4)$

d A which divides CD in the ratio 1:5 where C is $(-2, 1, 3)$ and D is $(7, -2, -12)$

8 The line joining $A(-1, 0, 0)$ and $B(-9, 4, 12)$ is divided by the point P in the ratio 1:3. Find the coordinates of the point P.

9 A line segment CD is divided in the ratio 3:1 by point F. The endpoints of the line are $C(3, -2, -5)$ and $D(1, 2, -1)$. Find the coordinates of F.

10 $P(5, 6, -6)$ and $R(-2, -15, 8)$ are two points. A divides PR in the ratio 2:5 and B divides PR in the ratio 5:2.

a Find the coordinates of A and B.

b Find the components of $\overrightarrow{AB}$ in terms of **i**, **j** and **k**.

Exercise 5.4 Vector pathways, parallel vectors and collinearity

Use the diagram of a cuboid for Questions 1 and 2.

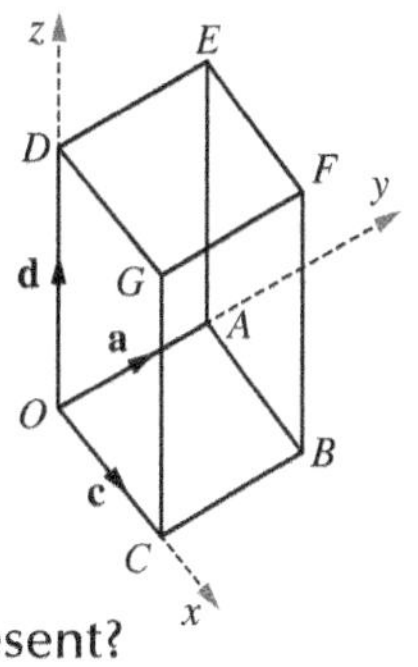

1 a $\overrightarrow{OA}$ represents the vector **a**.
Find three other representatives for **a**.

b $\overrightarrow{OC}$ represents the vector **c**.
Find three other representatives for **c**.

c $\overrightarrow{OD}$ represents the vector **d**.
Find three other representatives for **d**.

2 In terms of the vectors **a**, **c** and **d** what do these represent?

a $\overrightarrow{OB}$ b $\overrightarrow{DF}$ c $\overrightarrow{OE}$ d $\overrightarrow{AF}$ e $\overrightarrow{CO}$ f $\overrightarrow{FG}$

g $\overrightarrow{BA}$ h $\overrightarrow{AD}$ i $\overrightarrow{GB}$ j $\overrightarrow{OF}$ k $\overrightarrow{CE}$ l $\overrightarrow{BD}$

Example

Show that $P(-4, 6, 0)$, $Q(0, 4, 2)$ and $R(2, 3, 3)$ are collinear and find the ratio in which Q divides PR.

$$\overrightarrow{PQ} = \mathbf{q} - \mathbf{p} = \begin{pmatrix} 0 \\ 4 \\ 2 \end{pmatrix} - \begin{pmatrix} -4 \\ 6 \\ 0 \end{pmatrix} = \begin{pmatrix} 4 \\ -2 \\ 2 \end{pmatrix}$$

Calculate the components of $\overrightarrow{PQ}$.

$$\overrightarrow{QR} = \mathbf{r} - \mathbf{q} = \begin{pmatrix} 2 \\ 3 \\ 3 \end{pmatrix} - \begin{pmatrix} 0 \\ 4 \\ 2 \end{pmatrix} = \begin{pmatrix} 2 \\ -1 \\ 1 \end{pmatrix}$$

Calculate the components of $\overrightarrow{QR}$.

Note: You may not need to use position vectors.

Notice that $\overrightarrow{PQ} = 2\overrightarrow{QR}$

One set of components will be a multiple of the other set.

So $PQ \parallel QR$

The lines PQ and QR are parallel.

Q is a common point so P, Q and R are collinear.

The common point must be identified and mentioned.

Now draw a quick sketch.

Q divides PR in the ratio 2:1

Write down the ratio (in the correct order).

3 Show that the three given points are collinear and in each case find the ratio in which the 'middle' point divides the line joining the endpoints (in the given order).

a $A(2, -1, 4)$, $B(3, 2, 5)$, $C(6, 11, 8)$

b $P(1, -1, 3)$, $Q(3, 4, 0)$, $R(7, 14, -6)$

c $S(2, -1, -1)$, $T(4, 3, -5)$, $U(5, 5, -7)$

d $V(-1, 2, -3)$, $W(2, -4, -3)$, $X(4, -8, -3)$

e $D(-1, 3, 5)$, $E(-3, 5, 1)$, $F(-6, 8, -5)$

f $J(4, -7, 10)$, $K(0, 1, -2)$, $L(-5, 11, -17)$

g $M(25, 25, -10)$, $N(5, 10, -5)$, $P(-3, 4, -3)$

h $G\left(-\frac{3}{2}, 0, \frac{7}{2}\right)$, $H\left(\frac{3}{2}, -1, \frac{3}{2}\right)$, $I\left(3, -\frac{3}{2}, \frac{1}{2}\right)$

6 Working with vectors

Exercise 6.1 Equilibrium problems

1 Each diagram shows a set of three vectors representing forces around a point. Determine in each case whether the forces are in equilibrium.

a $\mathbf{u} = \begin{pmatrix} 4 \\ -4 \\ 0 \end{pmatrix}$ $\mathbf{v} = \begin{pmatrix} -5 \\ 2 \\ 1 \end{pmatrix}$ $\mathbf{w} = \begin{pmatrix} 1 \\ 2 \\ -1 \end{pmatrix}$

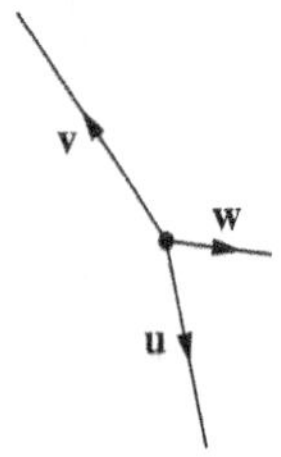

b $\mathbf{p} = \begin{pmatrix} 4 \\ -4 \\ 0 \end{pmatrix}$ $\mathbf{q} = \begin{pmatrix} -2 \\ 1 \\ 1 \end{pmatrix}$ $\mathbf{r} = \begin{pmatrix} -2 \\ 3 \\ 1 \end{pmatrix}$

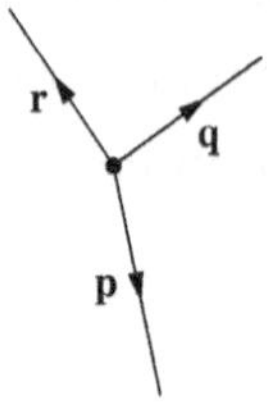

c

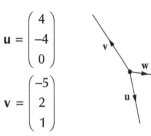

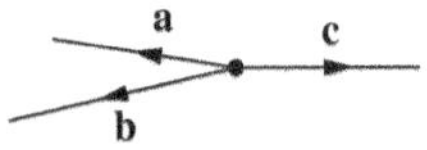

$\mathbf{a} = \begin{pmatrix} -8 \\ 3 \\ 2 \end{pmatrix}$ $\mathbf{b} = \begin{pmatrix} -6 \\ 5 \\ 4 \end{pmatrix}$ $\mathbf{c} = \begin{pmatrix} 14 \\ -8 \\ -6 \end{pmatrix}$

Hint A set of forces acting on a point are in equilibrium if the sum of the vectors representing the forces equals the zero vector.

2 Each set of three vectors represent forces that are in equilibrium around a point. In each case find the values of x, y and z.

a $\mathbf{f}_1 = \begin{pmatrix} x \\ 3 \\ 1 \end{pmatrix}$ $\mathbf{f}_2 = \begin{pmatrix} 4 \\ y \\ 2 \end{pmatrix}$ $\mathbf{f}_3 = \begin{pmatrix} 2 \\ -1 \\ z \end{pmatrix}$

b $\mathbf{f}_1 = \begin{pmatrix} -2 \\ -5 \\ 6 \end{pmatrix}$ $\mathbf{f}_2 = \begin{pmatrix} 5 \\ -4 \\ 0 \end{pmatrix}$ $\mathbf{f}_3 = \begin{pmatrix} x \\ y \\ z \end{pmatrix}$

c $\mathbf{f}_1 = \begin{pmatrix} x \\ -8 \\ -2 \end{pmatrix}$ $\mathbf{f}_2 = \begin{pmatrix} -1 \\ y \\ z \end{pmatrix}$ $\mathbf{f}_3 = \begin{pmatrix} 4 \\ -3 \\ 13 \end{pmatrix}$

d $\mathbf{f}_1 = \begin{pmatrix} 7 \\ 3 \\ z \end{pmatrix}$ $\mathbf{f}_2 = \begin{pmatrix} -4 \\ y \\ -6 \end{pmatrix}$ $\mathbf{f}_3 = \begin{pmatrix} x \\ -9 \\ 10 \end{pmatrix}$

3 Each set of three vectors represent forces that are in equilibrium around a point. In each case find the values of a, b and c.

a $\mathbf{f}_1 = a\mathbf{i} + 2\mathbf{j} - 3\mathbf{k}$ $\quad \mathbf{f}_2 = -2\mathbf{i} + b\mathbf{j} + 7\mathbf{k}$ $\quad \mathbf{f}_3 = \mathbf{i} - 3\mathbf{j} + c\mathbf{k}$

b $\mathbf{f}_1 = -2\mathbf{i} + 2\mathbf{j} - 3\mathbf{k}$ $\quad \mathbf{f}_2 = -5\mathbf{j} + 7\mathbf{k}$ $\quad \mathbf{f}_3 = a\mathbf{i} + b\mathbf{j} + c\mathbf{k}$

c $\mathbf{f}_1 = -\mathbf{i} + c\mathbf{k}$ $\quad \mathbf{f}_2 = a\mathbf{i} + b\mathbf{j}$ $\quad \mathbf{f}_3 = 4\mathbf{i} - 2\mathbf{j} - 2\mathbf{k}$

4 Three vectors are given by $\mathbf{p} = \begin{pmatrix} -1 \\ 2 \\ 0 \end{pmatrix}$, $\mathbf{q} = \begin{pmatrix} 0 \\ -4 \\ 3 \end{pmatrix}$ and $\mathbf{r} = \begin{pmatrix} 2 \\ 0 \\ -3 \end{pmatrix}$

$\mathbf{f}_1$, $\mathbf{f}_2$ and $\mathbf{f}_3$ represent three forces acting on a point.

In each case determine whether the system is in equilibrium.

a $\mathbf{f}_1 = 2\mathbf{p}$ $\quad \mathbf{f}_2 = \mathbf{q}$ $\quad \mathbf{f}_3 = \mathbf{r}$

b $\mathbf{f}_1 = \mathbf{q} - \mathbf{r}$ $\quad \mathbf{f}_2 = \mathbf{p} - \mathbf{r}$ $\quad \mathbf{f}_3 = \mathbf{p} - \mathbf{q}$

c $\mathbf{f}_1 = \mathbf{p}$ $\quad \mathbf{f}_2 = \mathbf{p} + \mathbf{q}$ $\quad \mathbf{f}_3 = \mathbf{r}$

d $\mathbf{f}_1 = 2\mathbf{p} - \mathbf{q}$ $\quad \mathbf{f}_2 = 2\mathbf{q}$ $\quad \mathbf{f}_3 = \mathbf{q} + \mathbf{r}$

e $\mathbf{f}_1 = 3\mathbf{p} + \mathbf{r}$ $\quad \mathbf{f}_2 = -\mathbf{p}$ $\quad \mathbf{f}_3 = \mathbf{q}$

f $\mathbf{f}_1 = 3\mathbf{p} + \mathbf{q}$ $\quad \mathbf{f}_2 = 2\mathbf{p} + \mathbf{r}$ $\quad \mathbf{f}_3 = -3\mathbf{p}$

Exercise 6.2 Scalar product

Reminder

Using magnitudes and the angle

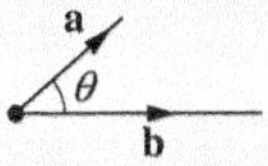

$\mathbf{a} \cdot \mathbf{b} = |\mathbf{a}|\,|\mathbf{b}| \cos\theta$

Using components

$\mathbf{a} = \begin{pmatrix} x_1 \\ y_1 \\ z_1 \end{pmatrix}$

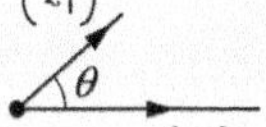

$\mathbf{b} = \begin{pmatrix} x_2 \\ y_2 \\ z_2 \end{pmatrix}$

$\mathbf{a} \cdot \mathbf{b} = x_1x_2 + y_1y_2 + z_1z_2$

These two different calculations give the same value for $\mathbf{a} \cdot \mathbf{b}$

1 In each case calculate $\mathbf{p} \cdot \mathbf{q}$ giving the exact value.

a $\mathbf{p} = \begin{pmatrix} -1 \\ 2 \\ 0 \end{pmatrix}, \mathbf{q} = \begin{pmatrix} 2 \\ -1 \\ 3 \end{pmatrix}$

b $\mathbf{p} = \begin{pmatrix} 1 \\ 3 \\ 1 \end{pmatrix}, \mathbf{q} = \begin{pmatrix} -2 \\ 0 \\ 1 \end{pmatrix}$

c $\mathbf{p} = \begin{pmatrix} 11 \\ -2 \\ -1 \end{pmatrix}, \mathbf{q} = \begin{pmatrix} 1 \\ -1 \\ -1 \end{pmatrix}$

d $\mathbf{p} = \begin{pmatrix} \frac{1}{2} \\ \frac{1}{2} \\ 2 \end{pmatrix}, \mathbf{q} = \begin{pmatrix} -4 \\ 6 \\ \frac{1}{2} \end{pmatrix}$

e $\mathbf{p} = \begin{pmatrix} \sqrt{2} \\ -\sqrt{2} \\ -1 \end{pmatrix}, \mathbf{q} = \begin{pmatrix} 1 \\ -1 \\ -\sqrt{2} \end{pmatrix}$

f $\mathbf{p} = \begin{pmatrix} \sqrt{3} \\ 2\sqrt{3} \\ -\sqrt{3} \end{pmatrix}, \mathbf{q} = \begin{pmatrix} \sqrt{3} \\ \sqrt{3} \\ 3\sqrt{3} \end{pmatrix}$

2 $\mathbf{v} = \begin{pmatrix} 0 \\ 1 \\ 1 \end{pmatrix}, \mathbf{w} = \begin{pmatrix} -1 \\ 0 \\ 1 \end{pmatrix}, |\mathbf{v}| = \sqrt{2}, |\mathbf{w}| = \sqrt{2}$ and $\theta = 60$.

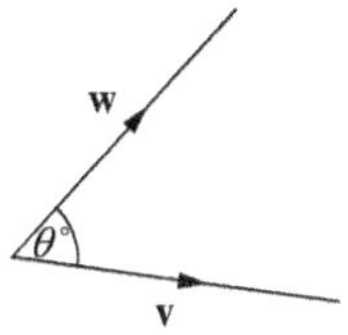

a Use the components to calculate $\mathbf{v} \cdot \mathbf{w}$

b Use the magnitudes and angle to calculate $\mathbf{v} \cdot \mathbf{w}$

c What do you notice?

3 $\mathbf{p} = \begin{pmatrix} 2 \\ 2 \\ 0 \end{pmatrix}, \mathbf{q} = \begin{pmatrix} 1 \\ 2 \\ 1 \end{pmatrix}, |\mathbf{p}| = 2\sqrt{2}, |\mathbf{q}| = \sqrt{6}$ and $\theta = 30$.

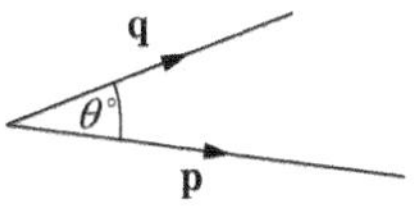

a Use the components to calculate $\mathbf{p} \cdot \mathbf{q}$

b Use the magnitudes and angle to calculate $\mathbf{p} \cdot \mathbf{q}$

c What do you notice?

4 In each case find an expression for $\mathbf{r} \cdot \mathbf{s}$ in terms of k.

a $\mathbf{r} = \begin{pmatrix} k \\ 2k \\ k \end{pmatrix}, \mathbf{s} = \begin{pmatrix} 3 \\ -2 \\ 2 \end{pmatrix}$

b $\mathbf{r} = \begin{pmatrix} -1 \\ k \\ -k \end{pmatrix}, \mathbf{s} = \begin{pmatrix} -k \\ 2 \\ 4 \end{pmatrix}$

c $\mathbf{r} = \begin{pmatrix} 2 \\ k+1 \\ k-1 \end{pmatrix}, \mathbf{s} = \begin{pmatrix} k-2 \\ 3 \\ -1 \end{pmatrix}$

d $\mathbf{r} = \begin{pmatrix} 2k+1 \\ 3k \\ -2 \end{pmatrix}, \mathbf{s} = \begin{pmatrix} -1 \\ 4 \\ k-2 \end{pmatrix}$

5 In each case calculate the value of p.

a $\begin{pmatrix} p \\ 1 \\ 3 \end{pmatrix} \cdot \begin{pmatrix} 2 \\ -1 \\ -1 \end{pmatrix} = 0$ **b** $\begin{pmatrix} 2p+1 \\ p-1 \\ 1 \end{pmatrix} \cdot \begin{pmatrix} 3 \\ -1 \\ -p \end{pmatrix} = 0$

c $(p\mathbf{i} - \mathbf{j} + 3\mathbf{k}) \cdot (2\mathbf{i} + p\mathbf{j} - \mathbf{k}) = 0$

Exercise 6.3 The angle between two vectors

Example

Calculate the size of angle DEF where D is (–1, 2, 3), E is (2, 5, 4) and F is (–1, 0, 3).

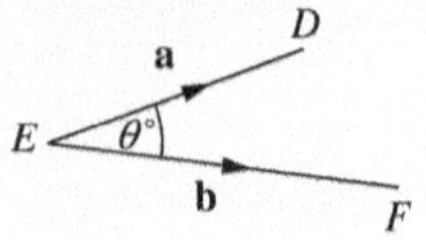

Sketch the angle DEF making sure the arrows point away from the vertex E. Label the vectors **a** and **b** and the angle $\theta°$.

$\mathbf{a} = \overrightarrow{DE} = \mathbf{e} - \mathbf{d} = \begin{pmatrix} 2 \\ 5 \\ 4 \end{pmatrix} - \begin{pmatrix} -1 \\ 2 \\ 3 \end{pmatrix} = \begin{pmatrix} 3 \\ 3 \\ 1 \end{pmatrix}$

$\mathbf{b} = \overrightarrow{DF} = \mathbf{f} - \mathbf{d} = \begin{pmatrix} -1 \\ 0 \\ 3 \end{pmatrix} - \begin{pmatrix} -1 \\ 2 \\ 3 \end{pmatrix} = \begin{pmatrix} 0 \\ -2 \\ 0 \end{pmatrix}$

Find the components of vectors **a** and **b**. You may manage this without using position vectors **d, e** and **f.**

$|\mathbf{a}| = \sqrt{3^2 + 3^2 + 1^2} = \sqrt{19}$

$|\mathbf{b}| = \sqrt{0^2 + (-2)^2 + 0^2} = 2$

Calculate the magnitudes of **a** and **b**.

$\mathbf{a} \cdot \mathbf{b} = \begin{pmatrix} 3 \\ 3 \\ 1 \end{pmatrix} \cdot \begin{pmatrix} 0 \\ -2 \\ 0 \end{pmatrix} = 3 \times 0 + 3 \times (-2) + 1 \times 0 = -6$

Calculate the scalar (dot) product.

$\cos\theta = \dfrac{\mathbf{a} \cdot \mathbf{b}}{|\mathbf{a}||\mathbf{b}|} = \dfrac{-6}{\sqrt{19} \times 2} = -0{\cdot}688\ldots$

Use the formula $\cos\theta = \dfrac{\mathbf{a} \cdot \mathbf{b}}{|\mathbf{a}||\mathbf{b}|}$

$\theta = 180 - \cos^{-1}(0{\cdot}688\ldots)$

$= 180 - 46{\cdot}5\ldots \approx 133{\cdot}5$

cos θ is negative so θ is a 2nd quadrant angle.

Use $\cos^{-1}$ with **positive** 0·688… to get the 1st quadrant angle.

2nd quadrant angle = 180° – (1st quadrant angle)

1 In each case calculate the value of θ.

a

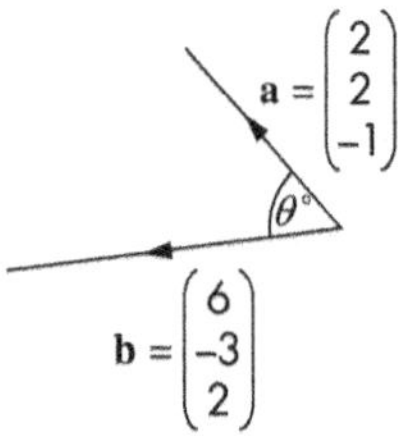

b

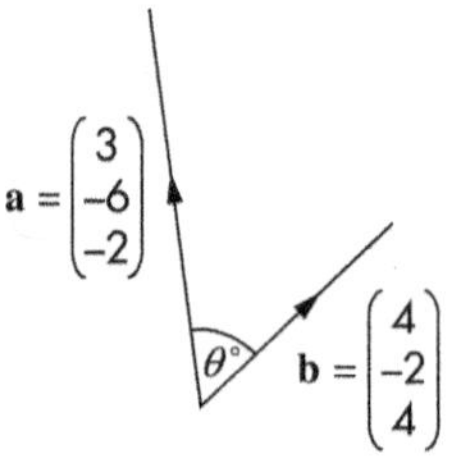

c

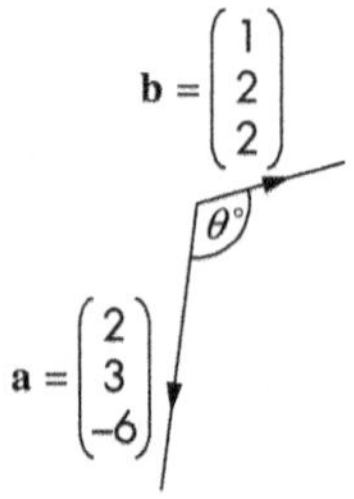

2 Use $\cos\theta° = \dfrac{\mathbf{a} \cdot \mathbf{b}}{|\mathbf{a}||\mathbf{b}|}$ to find θ.

a

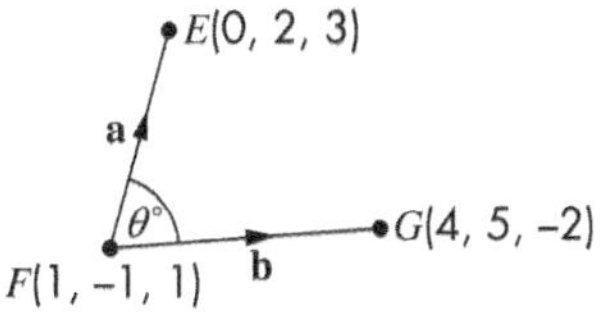

b

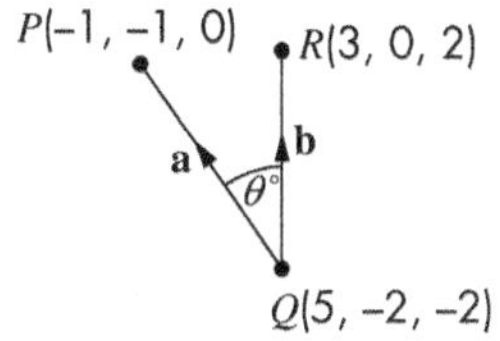

c

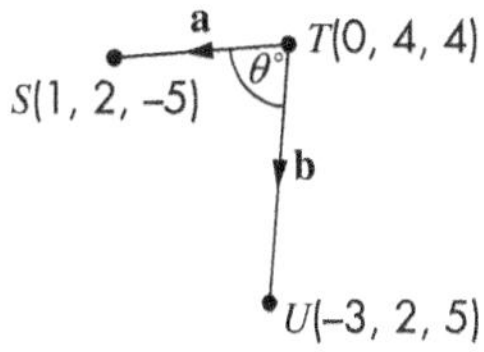

3 Find the size of angle *KLM* where *K* is (−2, 5, 4), *L* is (−1, 0, −3) and *M* is (2, 2, 8).

4 Find the sizes of the three angles of triangle *ABC* where *A* is (−1, 3, 5), *B* is (−2, −2, 4) and *C* is (0, 2, 3).

5 Calculate the sizes of the three angles of triangle *DEF*.

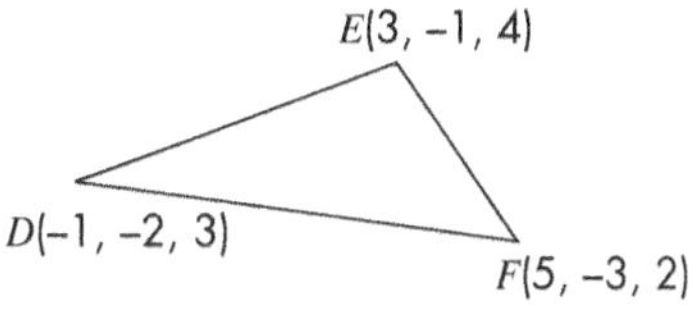

6 The diagram shows a cube at the origin with its edges along the axes.

The edge length of the cube is 1 unit so *C* has coordinates (1, 1, 1).

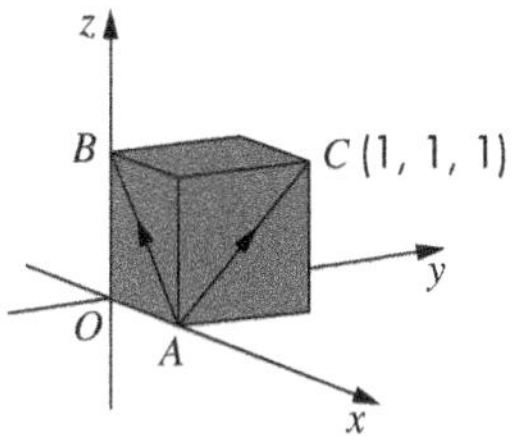

a Find the coordinates of points *A* and *B*.

b Find the components of $\overrightarrow{AB}$ and $\overrightarrow{AC}$.

c Calculate the size of angle *BAC*.

d By looking at triangle *ABC* can you explain your answer to part **c**?

7 Find the value of *k* for which the given vectors are perpendicular.

a $\begin{pmatrix} k \\ 2 \\ 3 \end{pmatrix}$ and $\begin{pmatrix} 1 \\ 1 \\ -1 \end{pmatrix}$ **b** $\begin{pmatrix} -1 \\ k \\ 3 \end{pmatrix}$ and $\begin{pmatrix} 2 \\ 1 \\ 0 \end{pmatrix}$ **c** $\begin{pmatrix} -1 \\ 3 \\ k+1 \end{pmatrix}$ and $\begin{pmatrix} 6 \\ -1 \\ 3 \end{pmatrix}$

d $\begin{pmatrix} -1 \\ -1 \\ -2 \end{pmatrix}$ and $\begin{pmatrix} 2 \\ k-2 \\ -1 \end{pmatrix}$

Hint If $\mathbf{a} \cdot \mathbf{b} = 0$ and **a** and **b** are non-zero vectors, then **a** and **b** are perpendicular.

Also if **a** and **b** are perpendicular then $\mathbf{a} \cdot \mathbf{b} = 0$

8 The vectors $\begin{pmatrix} m \\ -2 \\ -1 \end{pmatrix}$ and $\begin{pmatrix} m-1 \\ 1 \\ 4 \end{pmatrix}$ are perpendicular. Find the two possible values for *m*.

9 Show that triangle *ABC* is right-angled where *A* is (1, 3, 4), *B* is (3, −1, 0) and *C* is (3, 2, 6).

10 Show that triangle *PQR* is right-angled where *P* is (3, 7, −5), *Q* is (5, 9, −4) and *R* is (7, 5, −9) and calculate its area.

11 *ABC* is a right-angled triangle as shown in the diagram, with vertices *A*(1, −2, −*k*), *B*(*k*, *k*, 0) and *C*(4, −3, 3 − *k*).

Find the value of *k*.

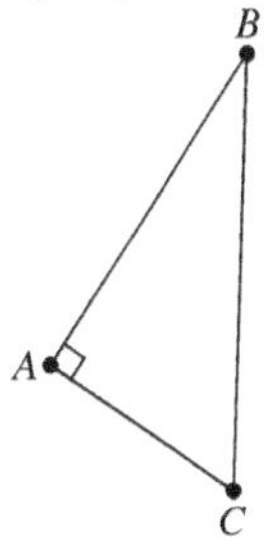

Exercise 6.4 Properties of the scalar product

1 In each case find the exact value of $\mathbf{a} \cdot \mathbf{b}$ (θ is the angle between **a** and **b**.)

a $|\mathbf{a}| = 2$, $|\mathbf{b}| = 3$ and $\theta° = 60°$ **b** $|\mathbf{a}| = 2$, $|\mathbf{b}| = 1$ and $\theta° = 30°$

c $|\mathbf{a}| = 5$, $|\mathbf{b}| = 10$ and $\theta = \frac{\pi}{3}$ **d** $|\mathbf{a}| = \sqrt{2}$, $|\mathbf{b}| = 1$ and $\theta = \frac{\pi}{4}$

2 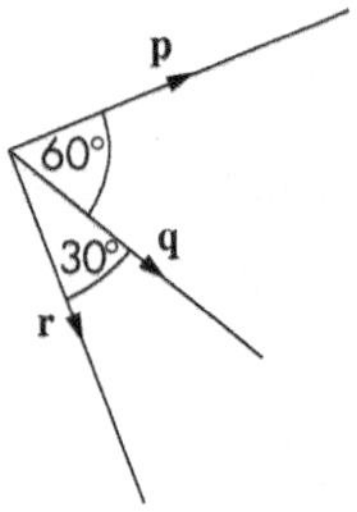

$|\mathbf{p}| = 3$, $|\mathbf{q}| = 2$ and $|\mathbf{r}| = 3$

Find the exact value of:

a $\mathbf{p} \cdot \mathbf{q}$ **b** $\mathbf{q} \cdot \mathbf{r}$ **c** $\mathbf{p} \cdot \mathbf{r}$

d $\mathbf{p} \cdot \mathbf{p}$ **e** $\mathbf{q} \cdot \mathbf{q}$ **f** $\mathbf{r} \cdot \mathbf{r}$

g $\mathbf{p} \cdot (\mathbf{q} + \mathbf{r})$ **h** $\mathbf{q} \cdot (\mathbf{p} + \mathbf{r})$

i $\mathbf{r} \cdot (\mathbf{q} - \mathbf{p})$ **j** $\mathbf{p} \cdot (\mathbf{p} - \mathbf{q})$

3 $|\mathbf{s}| = 1$, $|\mathbf{t}| = 2$ and $|\mathbf{u}| = 3$

Find the exact value of:

a $\mathbf{s} \cdot (\mathbf{t} + \mathbf{u})$ **b** $\mathbf{t} \cdot (\mathbf{u} - \mathbf{s})$

c $\mathbf{u} \cdot (\mathbf{u} + \mathbf{s})$ **d** $(\mathbf{t} + \mathbf{u}) \cdot (\mathbf{t} + \mathbf{u})$

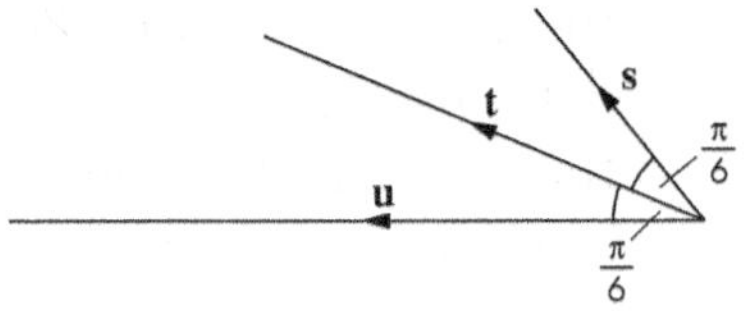

4 All six edges of this tetrahedron have length 2 units.

Evaluate $\mathbf{p} \cdot (\mathbf{q} + \mathbf{r})$

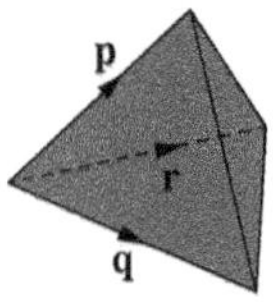

5 Each edge of this cube has length 1 unit.

a Find the exact value of $|\mathbf{b}|$ and $|\mathbf{c}|$.

b Find the exact value of $\mathbf{a} \cdot (\mathbf{b} + \mathbf{c})$

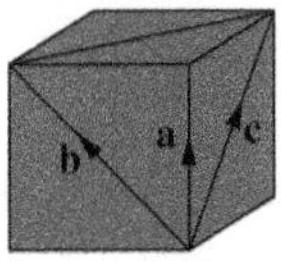

6 This prism has three squares faces and two that are equilateral triangles.
All the edges have length 4 units.

Evaluate $\mathbf{r} \cdot (\mathbf{s} + \mathbf{t})$

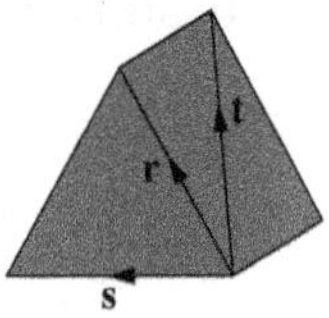

7 Solving algebraic equations

Exercise 7.1 Factorising polynomials

Example

Show that $x + 2$ is a factor of $f(x) = 2x^3 - 3x + 10$.

−2	2	0	−3	10
		−4	8	−10
	2	−4	5	0

Use −2 for the factor $x + 2$.

These are the coefficients of the polynomial; notice the 0 for the missing x^2 term.

The remainder is shown here: it is zero.

The remainder is 0 when $f(x)$ is divided by $x + 2$ so $x + 2$ is a factor of $f(x)$.

1 Show that in each case the given expression is a factor of $f(x)$.

a $x + 1; f(x) = x^3 - 6x^2 + 3x + 10$

b $x - 2; f(x) = x^3 - x^2 + x - 6$

c $x + 4; f(x) = 4x^3 + 16x^2 - x - 4$

d $x - 1; f(x) = x^3 - 7x + 6$

2 Show that $x + 2$ and $x - 4$ are both factors of $f(x) = x^4 - 15x^2 - 10x + 24$.

3 In each case determine whether or not the given expression is a factor of $f(x)$.

a $x - 3; f(x) = 2x^3 - x^2 + 3x - 1$

b $x + 1; f(x) = x^3 + 3x^2 - 2x - 4$

c $x + 2; f(x) = x^4 + 8x$

d $x - \frac{1}{2}; f(x) = 8x^3 + 4x^2 - 2x - 1$

Example

Factorise $x^3 + 2x^2 - 5x - 6$

Try $x - 1$ as a possible factor.

1	1	2	−5	−6
		1	3	−2
	1	3	−2	−8

The remainder is not zero so $x - 1$ is not a factor.

Now try $x + 1$ as a possible factor.

−1	1	2	−5	−6
		−1	−1	6
	1	1	−6	0

The remainder is zero so $x + 1$ is a factor.

So $f(x) = (x + 1)(x^2 + x - 6)$ — This is not factorised fully.

$= (x + 1)(x - 2)(x + 3)$ — This is factorised fully.

4 Show that the given expression is a factor of $f(x)$ and hence express $f(x)$ in fully factorised form.

a $x - 3; f(x) = x^3 - x^2 - 5x - 3$

b $x - 1; f(x) = x^3 - 6x^2 + 11x - 6$

c $x - 2; f(x) = x^3 - 7x + 6$

d $x - 1; f(x) = 6x^3 - 5x^2 - 2x + 1$

e $x + 2; f(x) = x^3 - 3x + 2$

f $x + 1; f(x) = 4x^3 + 4x^2 - x - 1$

g $x + 3; f(x) = 2x^3 + 7x^2 + 2x - 3$

h $x + 2; f(x) = x^3 - 2x^2 - 5x + 6$

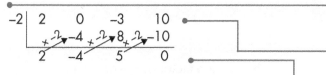

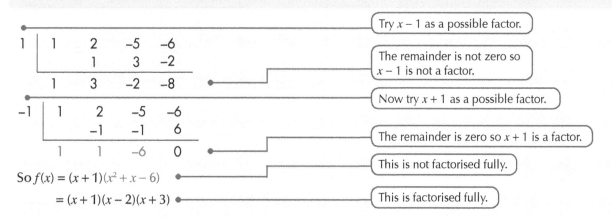

5 **a** Show that $x - 1$ is a factor of $f(x) = 2x^4 - 3x^3 - x^2 + 3x - 1$.

b Hence factorise $f(x)$ into two factors.

c Now show that $x + 1$ is a factor of one of these factors.

d Hence express $f(x)$ in fully factorised form.

6 In each case write the expression in fully factorised form. Show your working.

a $4x^3 - 4x^2 - x + 1$ **b** $x^3 - 2x - 1$ **c** $x^3 - 6x^2 + 12x - 8$

d $6x^3 + 13x^2 + x - 2$ **e** $2x^3 - 9x^2 + 8x + 3$ **f** $2x^3 - 3x + 10$

g $x^3 - 8$ **h** $x^3 + 27$ **i** $4x^3 - 4x^2 - 9x + 9$

7 Find the value of k given that:

a $x - 3$ is a factor of $x^3 - 2x^2 - 5x + k$ **b** $x + 1$ is a factor of $x^3 + kx^2 - x + 1$

c $x - 2$ is a factor of $x^3 + kx^2 + 4$ **d** $x - 1$ is a factor of $kx^3 - 3x^2 + 1$

e $x + 1$ is a factor of $x^4 + kx^3 - 3x^2 - 8x - 4$ **f** $x + 3$ is a factor of $2x^4 + 3x^3 + kx^2 - 7x + 6$

Exercise 7.2 Solving polynomial equations

Example

Solve the equation $x^3 - 5x^2 - 9x + 45 = 0$.

1	1	−5	−9	45
		1	−4	−13
	1	−4	−13	32

−1	1	−5	−9	45
		−1	6	3
	1	−6	−3	48

You are attempting to find factors of $x^3 - 5x^2 - 9x + 45$.

Try only factors of 45: 1, −1, 3, −3, 5, −5, etc.

3	1	−5	−9	45
		3	−6	−45
	1	−2	−15	0

Since the remainder on dividing by $x - 3$ is zero, then $x - 3$ is a factor.

So $x^3 - 5x^2 - 4x + 45 = (x - 3)(x^2 - 2x - 15)$

Use the bottom row: 1 −2 −15 to write down the quadratic factor.

$= (x - 3)(x - 5)(x + 3)$

Fully factorise by further factorising the quadratic factor $x^2 - 2x - 15$.

The equation becomes:

$(x - 3)(x - 5)(x + 3) = 0$

Rewrite the equation in factorised form.

$x - 3 = 0$ or $x - 5 = 0$ or $x + 3 = 0$

Set each factor to zero.

So $x = 3$ or $x = 5$ or $x = -3$

Solve for x.

The roots are −3, 3, 5

List the roots in increasing order.

1 Solve the following cubic equations.

a $x^3 - 3x + 2 = 0$ **b** $x^3 - 3x^2 - x + 3 = 0$

c $x^3 - 3x - 2 = 0$ **d** $x^3 - 8x^2 + 17x - 10 = 0$

e $x^3 - 2x^2 - 5x + 6 = 0$ **f** $x^3 + 4x^2 + x - 6 = 0$

g $x^3 - 3x^2 - 25x + 75 = 0$ **h** $x^3 - 5x^2 - 4x + 20 = 0$

2 Find the roots of these polynomial equations:

a $4x^3 - 8x^2 - x + 2 = 0$ **b** $3x^3 - x^2 - 3x + 1 = 0$

c $3x^3 - 2x^2 - 12x + 8 = 0$ **d** $4x^3 - 5x^2 - 7x + 2 = 0$

e $2x^3 - 3x^2 - 11x + 6 = 0$ **f** $5x^3 - 11x^2 + 7x - 1 = 0$

g $3x^3 - 11x^2 - 19x - 5 = 0$ **h** $4x^3 + 15x^2 + 17x + 6 = 0$

3 Solve these equations:

a $x^4 + x = 0$ **b** $y^3 - 7y + 6 = 0$ **c** $2x^3 - 9x^2 + 7x + 6 = 0$

d $3t^5 - 6t^3 + 3t = 0$ **e** $6x^3 - 7x^2 - x + 2 = 0$ **f** $4y^3 - 12y^2 - y + 3 = 0$

g $2x^4 - 12x^2 + 18x = 0$ **h** $x^3 + 4x^2 - 19x + 14 = 0$ **i** $6x^3 - 17x^2 - 4x + 3 = 0$

4 Solve these polynomial equations:

a $x^3 + x^2 = 8x + 12$ **b** $2x^3 + x = x^3 + 4x^2 - 6$

c $x^3 + x^2 - x = 6x^2 - 3x - 8$ **d** $2x^3 - 15x = x^3 + 2x^2$

e $x^4 + x^3 = x^2 + x$ **f** $x^4 + x^3 - 10x^2 = x^3 + 3x^2 - 36$

Exercise 7.3 Applications of solving polynomial equations

1 Find the x-axis and y-axis intercepts for the graphs of the following functions.

a $f(x) = (x + 1)(x - 1)(x - 2)$ **b** $g(x) = (x + 2)(x - 4)(x - 5)$

c $h(x) = x^3 - 5x^2 + 3x + 9$ **d** $f(x) = x^3 - 2x^2 - 11x + 12$

e $g(x) = x^3 - x^2 - 25x + 25$ **f** $h(x) = x^3 - 5x^2 - 2x + 24$

g $h(x) = x^3 + x^2 - 6x$ **h** $f(x) = x^3 + 10x^2 + 25x$

Hint To find the x-axis intercepts, solve $f(x) = 0$. To find the y-axis intercept, set $x = 0$.

2 Each diagram shows the graph of a cubic function $y = f(x)$.
In each case find a possible formula for $f(x)$ and determine the y-intercept.

a

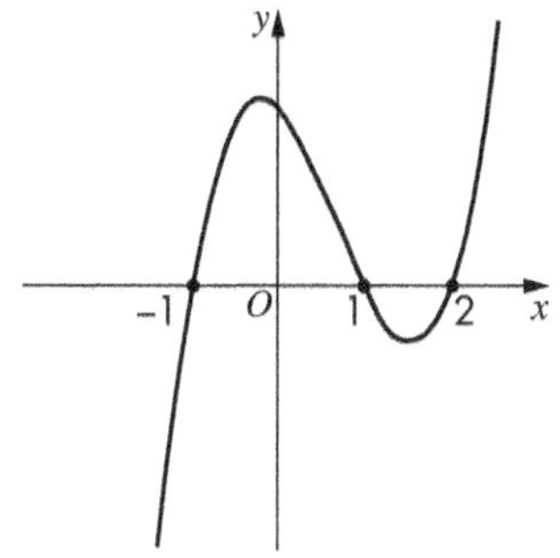

b

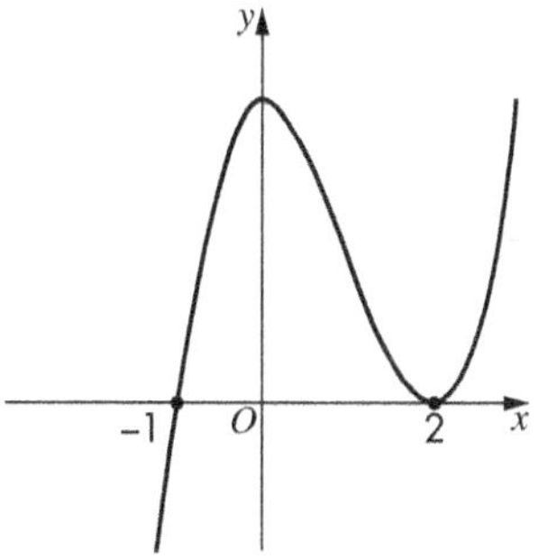

c

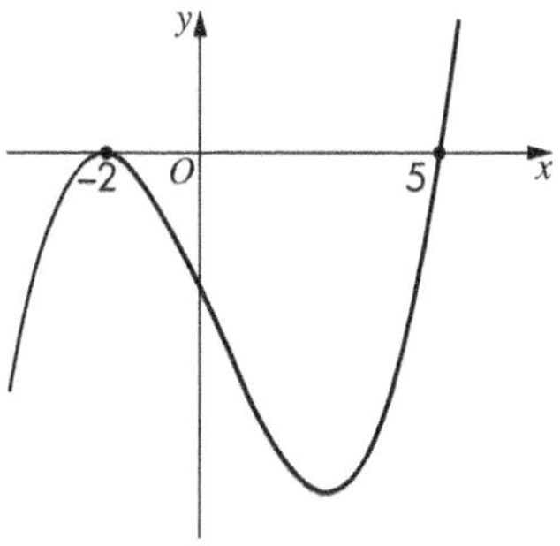

d

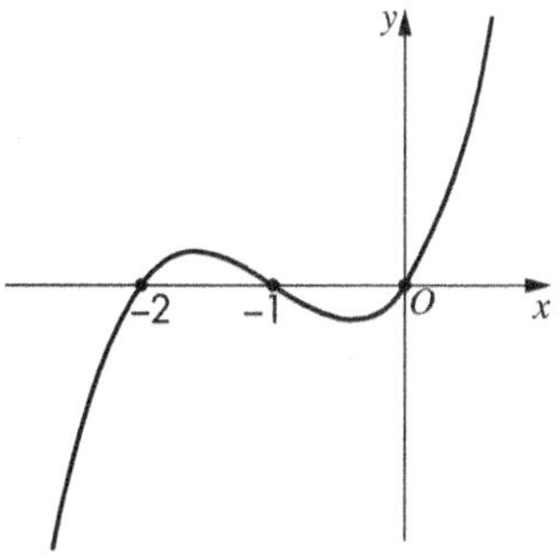

3 For each line find the coordinates of its points of intersection with the cubic graph $y = x^3 - x^2 - 2x - 1$.

a $y = 3x + 2$ **b** $y = -x - 2$ **c** $y = 4x - 1$

d $y = -1$ **e** $y = 2x - 5$

Hint Find the points of intersection of two graphs by setting their equations equal to each other.
Find the y-value by putting the x-value into one of the equations.

4 Find the coordinates of the points of intersection of the following lines and curves.

a $y = 2x + 3;\ y = x^3 - 5x^2 + 10x - 1$ **b** $y = 3x - 1;\ y = x^3 + x^2 - x - 5$

c $y = -2x + 10;\ y = x^3 + 4x^2 - 5x - 8$ **d** $y = x + 2;\ y = x^3 + 2x^2 - 8x - 16$

5 The diagram shows part of the graphs $y = x^3$ and $y = x^2 + 4x - 4$.
Find the coordinates of the three points of intersection.
(Note: the x-axis and y-axis scales are different.)

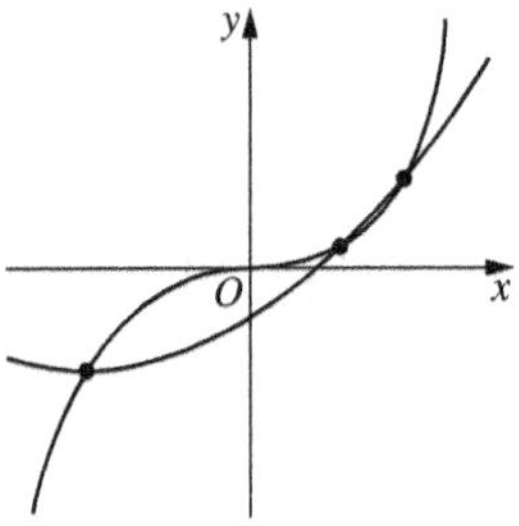

6 For each graph:

i find the x-axis intercepts

ii calculate the value of k.

a $y = k(x + 1)(x - 2)(x - 6)$ **b** $y = k(x + 3)(x + 1)(x - 2)$

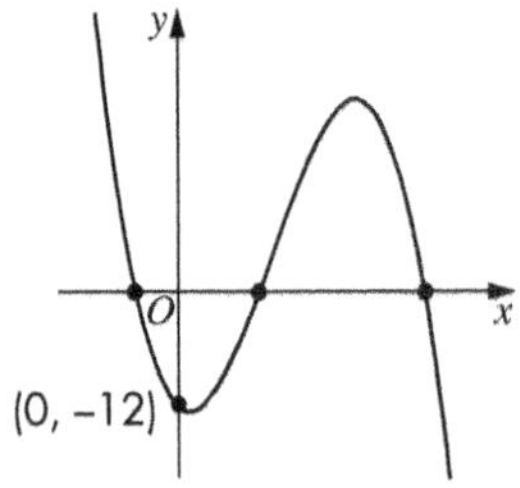

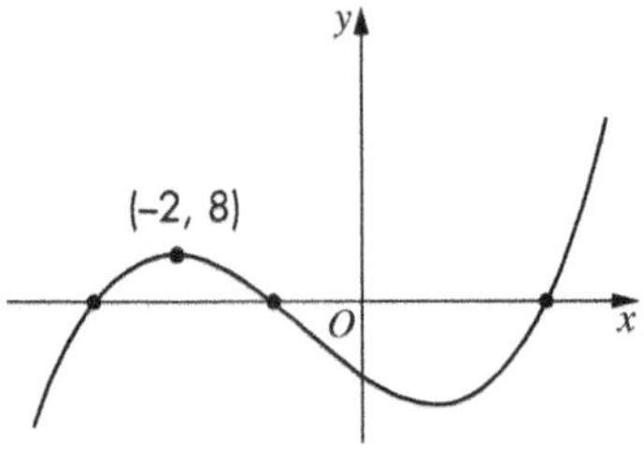

Hint Put the x- and y-coordinate values of the given point into the equation of the graph.

Exercise 7.4 Using the discriminant

Reminder

The graph of $y = ax^2 + bx + c$ meets the x-axis at 2 points.	The graph of $y = ax^2 + bx + c$ meets the x-axis at 1 point.	The graph of $y = ax^2 + bx + c$ does not meet the x-axis
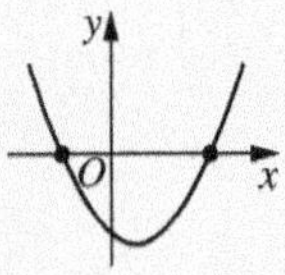	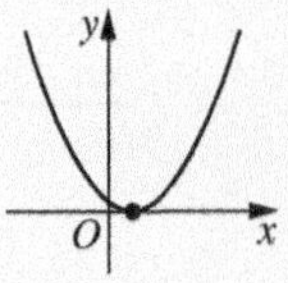	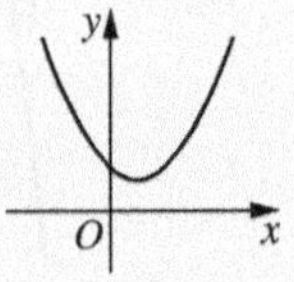
$ax^2 + bx + c = 0$ has **2 distinct real roots**	$ax^2 + bx + c = 0$ has **only 1 real root** (equal roots)	$ax^2 + bx + c = 0$ has **no real roots**
$b^2 - 4ac > 0$ **discriminant is positive**	$b^2 - 4ac = 0$ **discriminant is zero**	$b^2 - 4ac < 0$ **discriminant is negative**

Calculating the discriminant allows you to determine the **nature** of the roots:

$b^2 - 4ac > 0$	discriminant positive →	two distinct real roots
$b^2 - 4ac = 0$	discriminant zero →	one repeated real root (two equal real roots)
$b^2 - 4ac < 0$	discriminant negative →	no real roots

1 Sketch the following graphs showing clearly the x-intercepts.

a $y = x^2 - 3x$ **b** $y = x^2 + 4x - 5$ **c** $y = x^2 + 2x$

d $y = 5 + 4x - x^2$ **e** $y = 4x^2 - 1$ **f** $y = 2 + 3x - 2x^2$

g $y = 4x^2 + 4x - 15$ **h** $y = -2x^2 - 7x$ **i** $y = 6x^2 - 5x + 6$

> **Hint** Set $y = 0$ for x-intercepts.
>
> Positive x^2 Negative x^2
>
> 

2 Use the discriminant to determine the nature of the roots of these quadratic equations:

a $2x^2 + x + 1 = 0$ **b** $2x^2 + x - 1 = 0$ **c** $4x^2 - 4x + 1 = 0$

d $x^2 + 6x + 9 = 0$ **e** $x^2 - x + 6 = 0$ **f** $3x^2 + 3x - 1 = 0$

g $9x^2 - 12x + 4 = 0$ **h** $x^2 - 11x - 31 = 0$ **i** $x^2 + 4 = 0$

3 Which of these equations have equal real roots?

a $x^2 - 2x + 1 = 0$ **b** $4x^2 + 4x + 1 = 0$ **c** $4x^2 - 6x + 2 = 0$

d $x^2 + 8x + 16 = 0$ **e** $9x^2 - 6x - 1 = 0$ **f** $4x^2 + 8x + 3 = 0$

g $25x^2 - 40x + 16 = 0$ **h** $9x^2 + 24x + 16 = 0$ **i** $x^2 + x + \frac{1}{4} = 0$

4 Use the discriminant to show that the given line is a tangent to the curve.

a $y = 2x - 9$ and $y = 4x^2 + 14x$ **b** $y = 3x + 2$ and $y = x^2 - 11x + 51$

c $y = 10x - 1$ and $y = 25x^2$ **d** $y + 2x = 5$ and $x^2 + y^2 = 5$

e $x + 2y = 7$ and $x^2 + y^2 - 2x + 4y - 15 = 0$

> **Hint** A tangent to a curve has only one point of intersection. Set the equations equal or substitute for x or y, then show there is only one repeated root.

5 These equations have equal real roots. Find the value of a, b, or c in each case.

a $ax^2 + 4x + 2 = 0$ **b** $ax^2 + x + 1 = 0$ **c** $ax^2 - 3x + 3 = 0$

d $3x^2 - 2x + c = 0$ **e** $2x^2 + 4x + c = 0$ **f** $-3x^2 - 6x + c = 0$

g $2x^2 + bx + 8 = 0$ **h** $-x^2 + bx - 4 = 0$ **i** $x^2 + ax + a = 0$

> **Hint** There are two possibilities for parts **g**, **h** and **i**.

6 These equations have equal real roots. Find the possible values for k in each case.

a $kx^2 + kx + 2 = 0$ **b** $x^2 + (k - 1)x - k = 0$

c $(2k + 1)x^2 + (2k + 1)x + 1 = 0$ **d** $x(x + k) = -4$

e $x(x - k) = k$ **f** $(x + k)^2 + x^2 = 2$

g $k - (2x - k)^2 = 2$ **h** $(x - k)^2 + (2x - k)^2 = 5$

7 Solve the following inequalities using the graphs.

a $x^2 - x - 6 \leqslant 0$

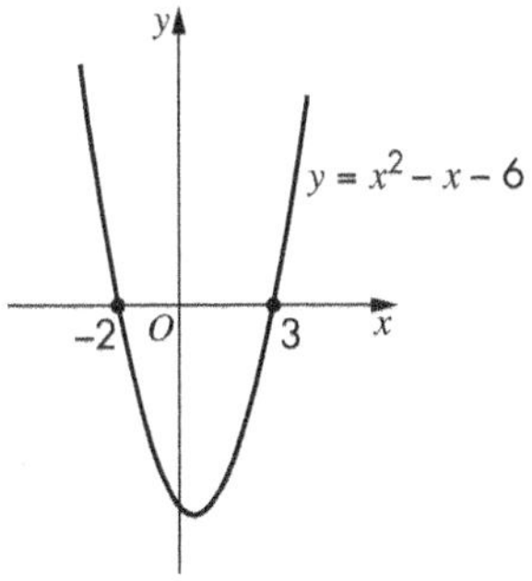

b $10 - 3x - x^2 > 0$

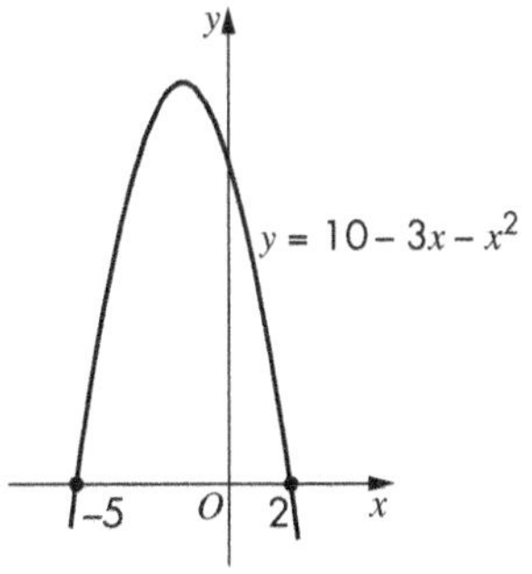

c $x^2 + 8x > 0$

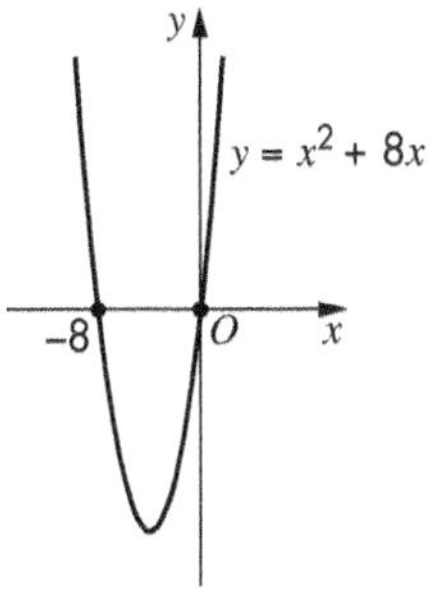

d $-x^2 - 4x - 3 \leqslant 0$

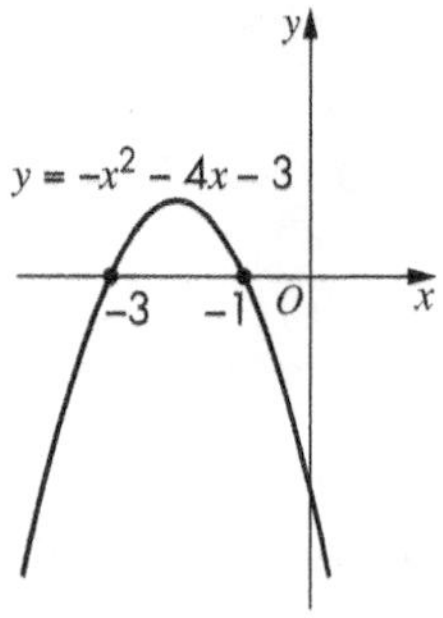

e $6 + 5x - x^2 \geqslant 0$

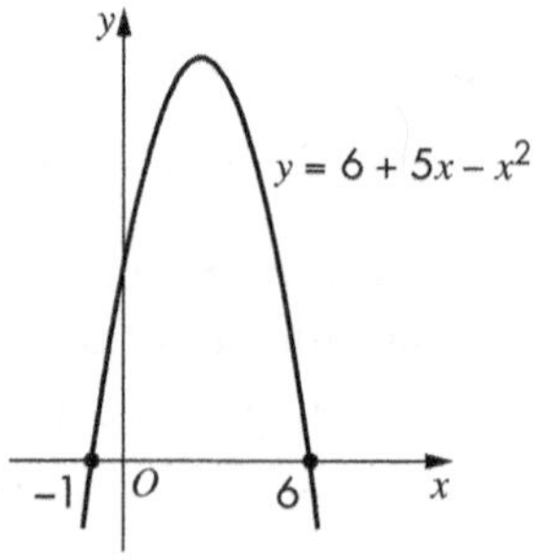

f $x^2 + 5x + 6 \geqslant 0$

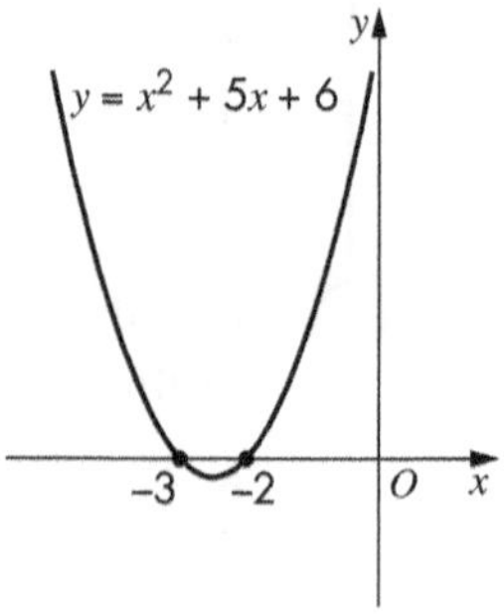

8 For what values of k does:

a $4x^2 + 8x + k = 0$ have real roots

b $kx^2 - 2x + 3 = 0$ have no real roots

c $2x^2 + kx + 2 = 0$ have equal real roots

d $kx^2 - 3x + k = 0$ have real roots

e $\dfrac{4(x-1)}{x^2+3} = k$ have equal real roots?

Exercise 7.5 Applications of solving logarithmic and exponential equations

> **Hint**
> **For Questions 1–8:**
> - substitute the given values into the formula
> - rearrange to make a 'power statement'
> - rewrite the 'power statement' as a 'log statement'
> - solve the equation.

1 A chemist investigated the cooling rates of various substances under different conditions.

The formula she used was $T_t = T_0 e^{-kt}$ ...where T_0 is the initial temperature (in °C) of the substance and T_t (in °C) is its temperature after t minutes.

The data that she gathered is shown in the table below. Calculate in each case the value of k to 3 significant figures.

	Substance	Initial temp. (°C)	Final temp. (°C)	Time elapsed
a	Propylamine	40	35	10 min
b	Butylamine	75	65	$\frac{1}{4}$ hour
c	Phenylamine	160	130	12 min
d	Triethylamine	80	79	1 min
e	Methanoic acid	75	64	20 min
f	Ethanoic acid	100	91	18 min
g	Propanoic acid	141	95	$\frac{1}{2}$ hour
h	Sorbic acid	228	200	9 min

2 The amount A grams of a radioactive substance after t units of time is given by $A = A_0 e^{-kt}$ where A_0 is the initial amount of the substance and k is a constant.

a After 8 days, 100 grams of thorium reduces to 74 grams. Use this information to calculate the value of k and hence calculate the half-life of thorium (the length of time, t, after which half the initial amount of the thorium is left).

b Calculate the half-life of neptunium if it takes 3 hours for 50 grams to reduce to 45·5 grams.

c Calculate the half-life of each of these isotopes from the given information:

i Plutonium has different isotopes. One of them is called plutonium-236. After 3 years, 436 grams of this isotope still remained out of 1000 grams.

ii For another isotope, plutonium-239, it is estimated that even after 1000 years, 972 grams of an original 1000 grams would still remain.

iii Tritium is the only radioactive form of hydrogen. It is used in the luminous dials of watches. 14 grams of this substance decays over the course of 2 years into 12·5 grams.

iv A man called Willard Libby developed a method of using the decay of carbon-14, a naturally occurring isotope of carbon, to date wood, seeds and bones. For example, knowing that 15 micrograms of carbon-14 will reduce to 11·8 micrograms over the course of 2000 years allows archaeologists to date samples that they discover.

3 Find the equation of the straight line passing through each pair of points.

a (2·3, 5·7), (3·3, 7·7) **b** (2, 3·5), (2·5, 3·75)

c (0·5, 2), (1·25, 0·5) **d** (0, 9·5), (0·2, 9·8)

4 Using a suitable pair of points, find the equation of each line in the form $Y = mX + c$

a

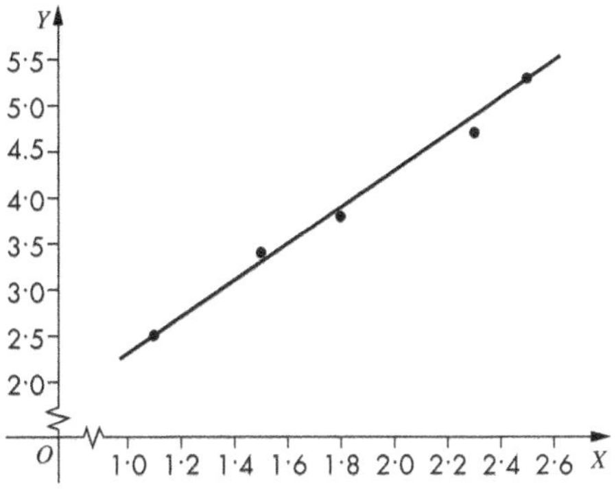

X	1·1	1·5	1·8	2·3	2·5
Y	2·5	3·4	3·8	4·7	5·3

b

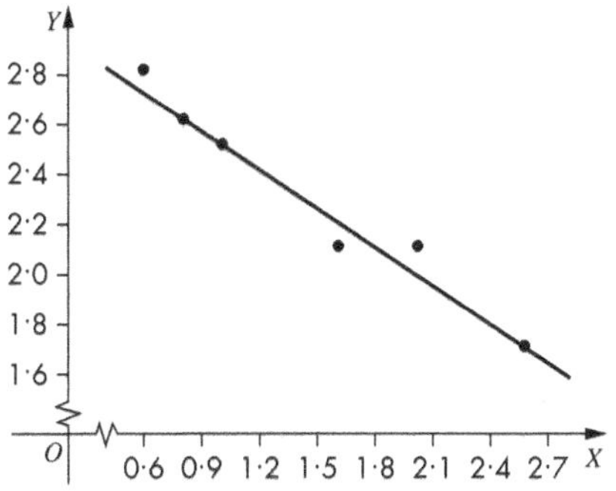

X	0·6	0·8	1·0	1·6	2·0	2·6
Y	2·8	2·6	2·5	2·1	2·1	1·7

5 It is known that x and y are connected by a relationship of the form $y = ax^b$. Find the values of the constants a and b given that:

a $\log_e y = 2\log_e x + 1{\cdot}5$ **b** $\log_e y = 3{\cdot}6\log_e x + 0{\cdot}8$ **c** $\log_e y = 0{\cdot}5\log_e x + 2{\cdot}3$

6 Experimental data is graphed and it is found that the straight line $B = 1{\cdot}2A + 0{\cdot}3$ fits the graph where $A = \log_e x$ and $B = \log_e y$.

It is also thought that x and y are connected by a relationship of the form $y = ax^b$. If this is true then find the values of the constants a and b.

7 The period of oscillation was measured for five different pendulums. It was thought that the length, x metres, of the pendulums and the period, y seconds, were connected by a relationship of the form $y = ax^b$.

After the measurements were taken the following table was constructed and a graph drawn.

$X (= \log_e x)$	−0·916	−0·511	−0·223	0	0·182
$Y (= \log_e y)$	0·223	0·430	0·561	0·680	0·772

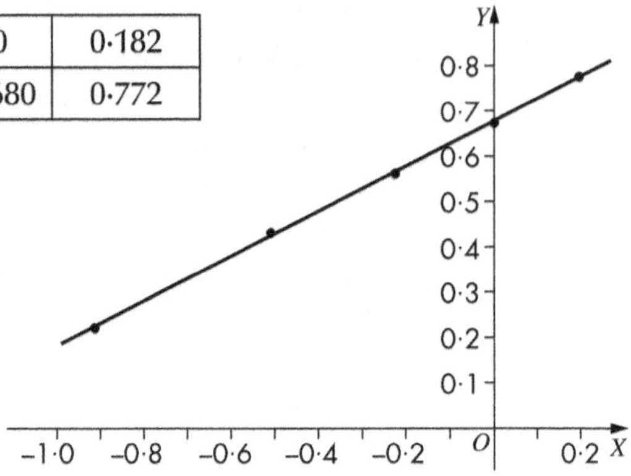

a Find the equation of the graph in the form $Y = mX + c$.

b Hence find the values of the constants a and b in the relationship $y = ax^b$.

8 The average distance, x units (taking $x = 1$ for the Earth), of a planet from the Sun and the time, y years, of one revolution round the Sun are thought to be connected by a relationship of the form $y = ax^b$. Using data for the five planets closest to the Sun in our own solar system the following table was constructed and a graph drawn.

Planet	Mercury	Venus	Earth	Mars	Jupiter
$X (= \log_e x)$	−0·949	−0·324	0	0·421	1·649
$Y (= \log_e y)$	−1·423	−0·485	0	0·632	2·473

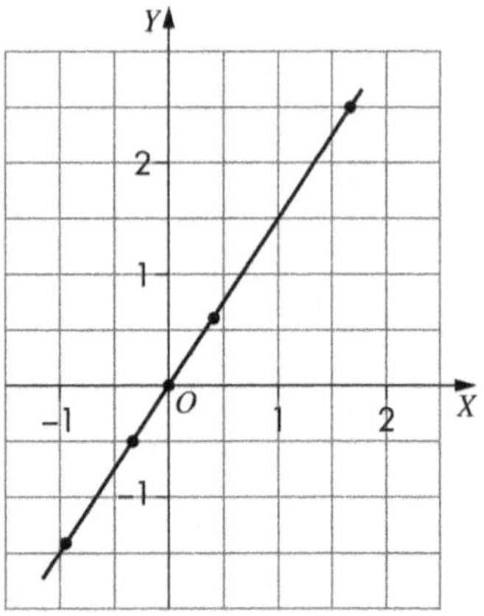

a Find the equation of the best-fitting line (shown in the diagram) in the form $Y = mX + c$.

b Hence find good estimates for the values of the constants a and b in the relationship $y = ax^b$.

c It is known that Uranus takes very close to 84 years to complete one revolution round the Sun. Use the relationship you discovered to estimate the distance of Uranus from the Sun.

8 Solving trigonometric equations

Exercise 8.1 Solving linear trigonometric equations

Reminder

This is the graph $y = \sin x°$

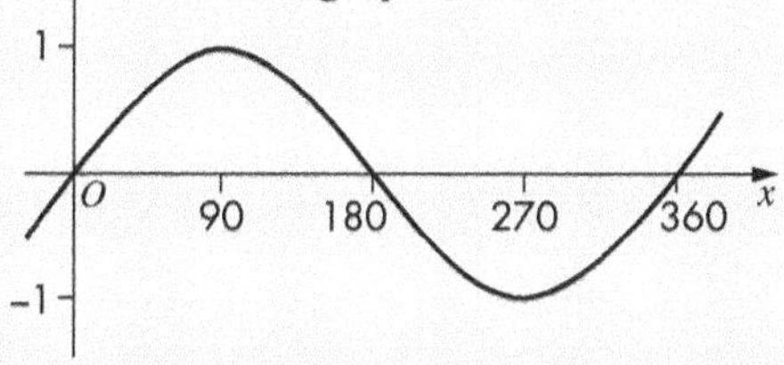

This is the graph $y = \cos x°$

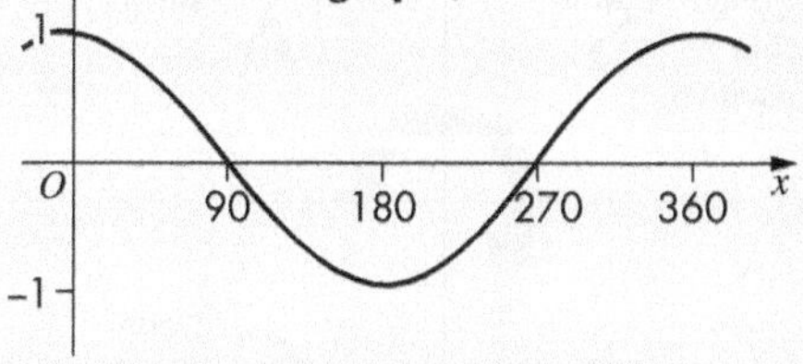

This is the graph $y = \sin x$

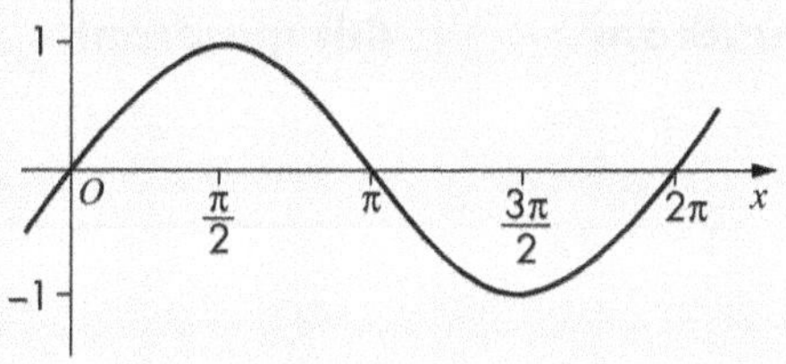

This is the graph $y = \cos x$

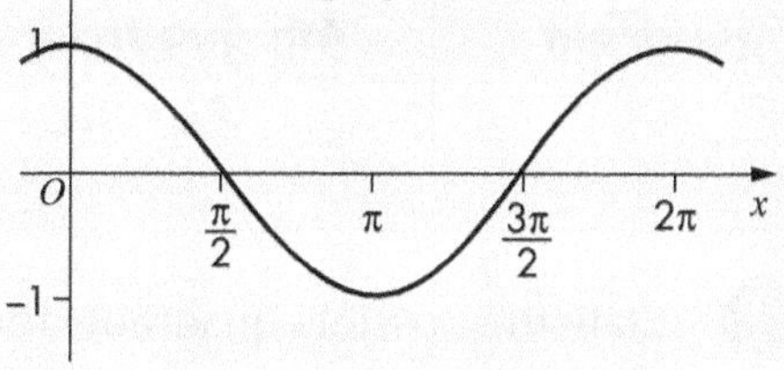

Each graph in the bottom row has a different x-axis scale (it is in radians) from the corresponding graph in the top row above (it is in degrees). So they are different graphs. The graph $y = \sin x°$ is different from the graph $y = \sin x$.

1 Solve:

a $\sin x° = 0$ for $0 \leqslant x \leqslant 360$

b $\sin x = 0$ for $0 \leqslant x \leqslant 2\pi$

c $\cos x° = 1$ for $0 \leqslant x < 360$

d $\cos x° = 1$ for $0 < x \leqslant 360$

e $\sin x = 0$ for $0 < x < 2\pi$

f $\sin x° = -1$ for $0 \leqslant x < 360$

g $\cos x = 1$ for $0 \leqslant x \leqslant 2\pi$

h $\cos x = -1$ for $0 \leqslant x \leqslant 2\pi$

i $\cos x° = 0$ for $180 \leqslant x < 360$

j $\sin x° = 1$ for $0 \leqslant x < 360$

k $\cos x° = -1$ for $0 \leqslant x \leqslant 360$

l $\cos x = 0$ for $0 \leqslant x \leqslant \pi$

Hint Read the inequality for x very carefully.

2 Find the exact value of:

a $\sin \frac{\pi}{4}$ b $\sin 60°$ c $\cos \frac{\pi}{3}$

d $\cos \frac{\pi}{4}$ e $\tan 45°$ f $\tan \frac{\pi}{3}$

g $\cos 30°$ h $\sin 30°$ i $\tan \frac{\pi}{4}$

j $\sin 45°$ k $\cos \frac{\pi}{6}$ l $\tan 30°$

Hint Use these diagrams:

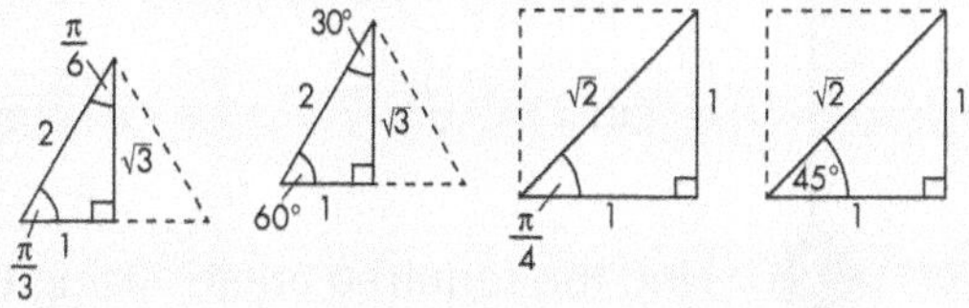

3 Find the value of x, where $0 < x < 90$ or $0 < x < \frac{\pi}{2}$ (i.e. 1st quadrant angle only). Take care – some are in degrees and some are in radians.

a $\sin x° = \frac{\sqrt{3}}{2}$

b $\tan x = \sqrt{3}$

c $\cos x = \frac{1}{2}$

d $\cos x° = \frac{\sqrt{3}}{2}$

e $\tan x° = \frac{1}{\sqrt{3}}$

f $\cos x° = \frac{1}{\sqrt{2}}$

g $2\sin x - \sqrt{3} = 0$

h $\sqrt{3}\tan x - 1 = 0$

Reminder

Which quadrants are the solutions in?

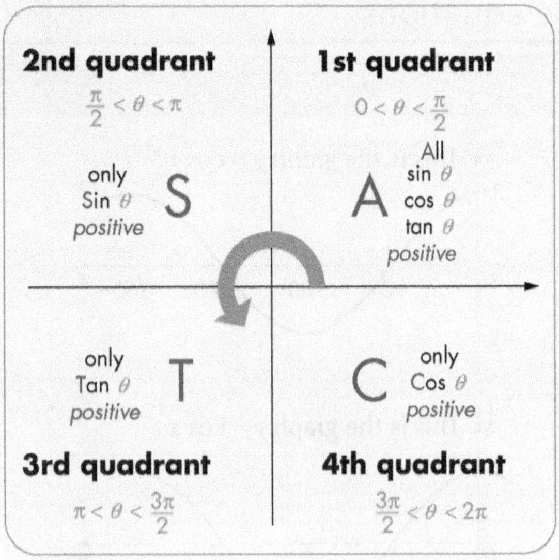

What to do with the 1st quadrant angle α?

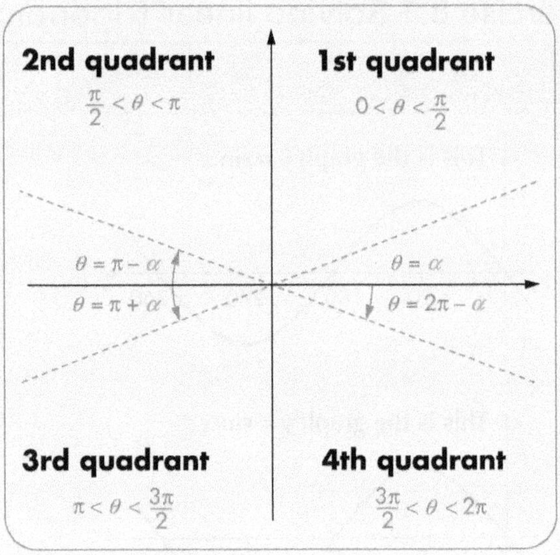

4 State the possible quadrants for angle x.

a $\sin x = 0{\cdot}2$ **b** $\tan x = -1$ **c** $\cos x = -\frac{1}{2}$

d $\sin x = -\frac{1}{\sqrt{2}}$ **e** $\cos x = \frac{\sqrt{3}}{2}$ **f** $\tan x = \frac{1}{\sqrt{3}}$

g $\sin x = \pm\frac{\sqrt{3}}{2}$ **h** $\cos x = \pm\frac{1}{\sqrt{2}}$ **i** $\tan x = \pm\sqrt{3}$

5 Find the 1st quadrant angle (to 1 d.p.) used when solving:

a $\sin x° = -\frac{1}{3}$ **b** $\cos x° = -0{\cdot}1$ **c** $\cos x° = \frac{2}{3}$

d $\tan x° = -\frac{3}{2}$ **e** $\sin x° = -\frac{\sqrt{3}}{7}$ **f** $\cos x° = -\frac{\sqrt{2}}{3}$

Hint Check your calculator is in DEG mode

6 Find the 1st quadrant angle (to 3 s.f.) used when solving:

a $\sin x = -0{\cdot}2$ **b** $\cos x = -0{\cdot}9$ **c** $\tan x = \frac{7}{5}$

d $\tan x = -\sqrt{3}$ **e** $\cos x = -\frac{1}{\sqrt{5}}$ **f** $\sin x = -\frac{2}{9}$

Hint Check your calculator is in RAD mode

7 Find the required angles using the given 1st quadrant angle.

	1st quadrant angle	Required angles
a	30°	3rd & 4th quadrants
b	60°	1st & 4th quadrants
c	45°	2nd & 3rd quadrants
d	23·6°	2nd & 4th quadrants
e	80·1°	all 4 quadrants
f	12·9°	3rd & 4th quadrants

8 Solve for $0 \leqslant x \leqslant 360$ (to 1 d.p. where necessary):

a $\sin x° = \frac{1}{3}$ b $\cos x° = -\frac{1}{2}$ c $\cos x° = \frac{2}{3}$ d $\sin x° = -\frac{1}{2}$

e $\sin x° = \frac{1}{2}$ f $\sin x° = -\frac{1}{4}$ g $\cos x° = -\frac{3}{4}$ h $\cos x° = -\frac{1}{5}$

9 Solve for $0 \leqslant x \leqslant 2\pi$ (to 3 s.f.):

a $\sin x = 0{\cdot}1$ b $\cos x = \frac{1}{3}$ c $\tan x = -\frac{1}{2}$

d $\sin x = -\frac{2}{3}$ e $\cos x = -0{\cdot}92$ f $\tan x = \sqrt{5}$

Hint: What mode is your calculator in?

10 Simplify the following radian expressions.

a $\pi + \frac{\pi}{4}$ b $\pi - \frac{\pi}{3}$ c $2\pi + \frac{\pi}{6}$ d $2\pi - \frac{\pi}{3}$

e $\pi - \frac{\pi}{4}$ f $2\pi - \frac{\pi}{4}$ g $\pi + \frac{\pi}{3}$ h $\pi + \frac{\pi}{6}$

i $2\pi - \frac{\pi}{6}$ j $2\pi + \frac{\pi}{4}$ k $\pi - \frac{\pi}{2}$ l $\pi + \frac{3\pi}{2}$

Hint: Think of π as:
3 lots of $\frac{\pi}{3}$
or
4 lots of $\frac{\pi}{4}$
or
6 lots of $\frac{\pi}{6}$

11 Solve for $0 \leqslant x \leqslant 2\pi$:

a $2\sin x - 1 = 0$ b $2\cos x - \sqrt{3} = 0$ c $1 + \sqrt{2}\cos x = 0$

d $2\sin x + 1 = 0$ e $2\sin x + \sqrt{3} = 0$ f $\tan x + \sqrt{3} = 0$

12 Solve for $0 \leqslant x \leqslant 360$ (to 1 d.p. where necessary):

a $3\sin x° + 1 = 0$ b $5\cos x° - 1 = 0$ c $3\tan x° + 4 = 0$

d $5\sin x° - \sqrt{2} = 0$ e $2 - 3\cos x° = 0$ f $\sqrt{5}\tan x° + 1 = 0$

13 Solve for $0 \leqslant x \leqslant 2\pi$ (to 3 s.f.):

a $3\sin x + 1 = 0$ b $3\cos x - 2 = 0$ c $\sqrt{3}\tan x + 2 = 0$

14 Find the exact solutions of these equations for $0 \leqslant x \leqslant 2\pi$:

a $\sin x = -\frac{\sqrt{3}}{2}$ b $\tan x = \sqrt{3}$ c $\cos x = -\frac{1}{\sqrt{2}}$ d $\sin x = -\frac{1}{2}$

e $\sqrt{3}\tan x = 1$ f $\sqrt{2}\sin x = -1$ g $2\cos x = -\sqrt{3}$ h $2\cos x + 1 = 0$

15 Solve for $0 \leqslant x \leqslant 2\pi$:

a $\sin 2x = 1$ b $\cos 2x = 0$ c $\sin 3x = -1$ d $\sin^2 2x = 1$

e $\cos^2 3x = 1$ f $\sin 2x = \frac{1}{2}$ g $\cos 2x = -\frac{1}{2}$ h $\tan^2 2x = 1$

16 For each diagram find algebraically (do not use a graphic calculator) the coordinates of P and Q, the points of intersection of the two graphs.

a

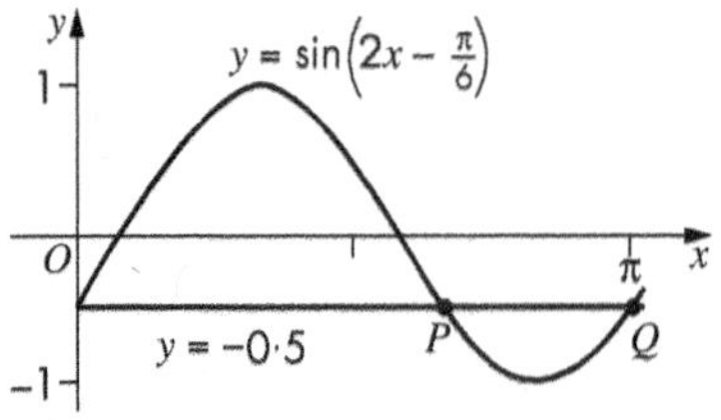

b

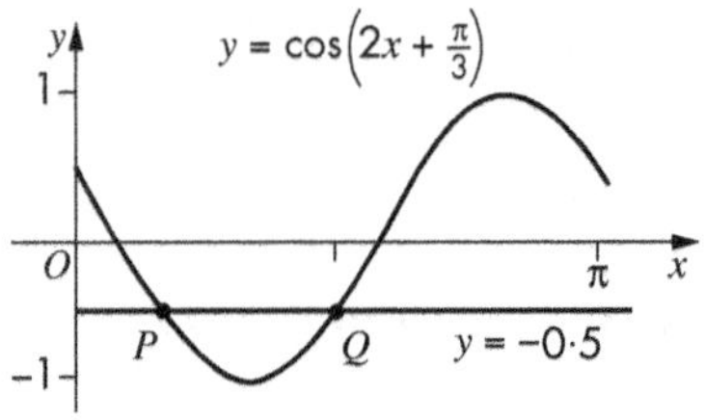

c

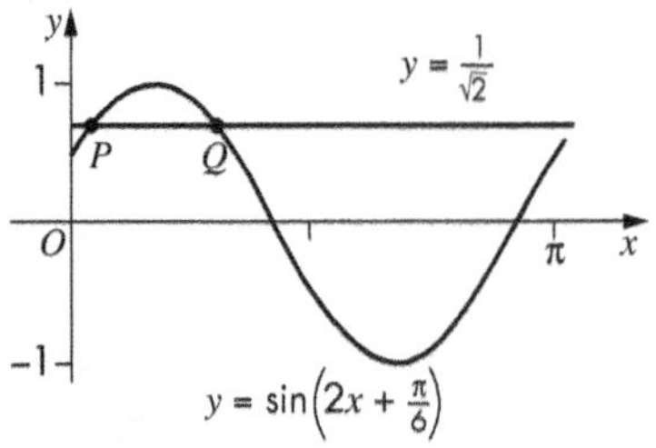

d

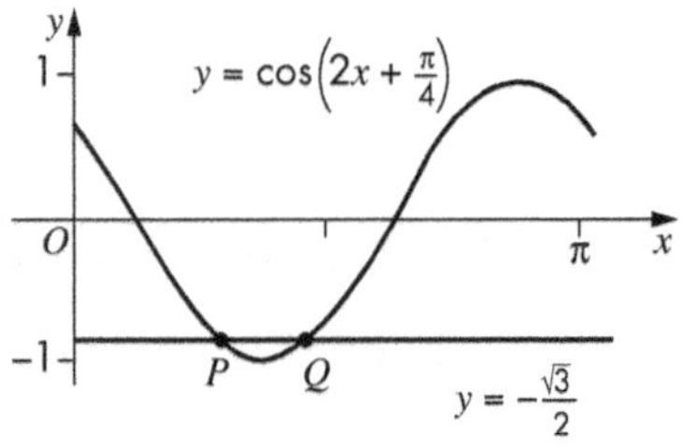

Exercise 8.2 Solving quadratic trigonometric equations

Reminder

Solving $x^2 = 4$ does not give only $x = \sqrt{4} = 2$, it also gives $x = -\sqrt{4} = -2$.

This is shortened to $\pm\sqrt{4} = \pm 2$.

So when solving the trig equation $\cos^2 x = \frac{1}{2}$ you get $\cos x = \pm\sqrt{\frac{1}{2}} = \pm\frac{1}{\sqrt{2}}$

This means $\cos^2 x = \frac{1}{2}$ gives two possibilities: $\cos x = \frac{1}{\sqrt{2}}$ or $\cos x = -\frac{1}{\sqrt{2}}$ and so angle x could be in any of the four quadrants.

1 Solve for θ where $0 \leqslant \theta \leqslant 360$:

a $\sin^2\theta° = 1$ **b** $\cos^2\theta° = 1$ **c** $\tan^2\theta° = 1$

d $\cos^2\theta° = \frac{1}{4}$ **e** $\tan^2\theta° = 3$ **f** $\sin^2\theta° = \frac{3}{4}$

2 Solve for x where $0 \leqslant x \leqslant 2\pi$:

a $\sin^2 x = \frac{1}{4}$ **b** $\sin^2 x = \frac{1}{2}$ **c** $\cos^2 x = \frac{1}{2}$

d $\cos^2 x = \frac{3}{4}$ **e** $\tan^2 x = \frac{1}{3}$ **f** $\sin^2 x = 1$

3 Solve for θ or x where $0 \leqslant \theta \leqslant 360$ or $0 \leqslant x \leqslant 2\pi$:

a $3\tan^2\theta° - 1 = 0$ **b** $4\cos^2 x - 1 = 0$ **c** $\tan^2 x - 2 = 1$

d $4\sin^2\theta° = 1$ **e** $\sin^2 x + 2 = 3$ **f** $2\sin^2\theta° - 1 = 0$

g $3\cos^2\theta° = 3 - \cos^2\theta°$ **h** $2\sin^2\theta° = 1 + \sin^2\theta°$

i $3\sin^2 x - 1 = 2 - \sin^2 x$ **j** $3\tan^2 x - 2 = 2\tan^2 x - 1$

4 Solve for θ where $0 \leqslant \theta \leqslant 360$ giving your answers correct to 1 decimal place:

a $\tan^2\theta° = 2$ **b** $\sin^2\theta° = 0{\cdot}7$ **c** $\cos^2\theta° = 0{\cdot}21$

d $3\sin^2\theta° = 1$ **e** $5\cos^2\theta° - 1 = 0$ **f** $2\tan^2\theta° - 7 = 0$

Hint What mode is your calculator in?

5 Solve for x where $0 \leqslant x \leqslant 2\pi$ giving your answers correct to 3 significant figures:

a $7\sin^2 x - 3 = \sin^2 x - 2$ **b** $4 - \cos^2 x = 5\cos^2 x + 3$

c $8 - 3\tan^2 x = 2\tan^2 x$ **d** $2{\cdot}8\sin^2 x = 5{\cdot}7\sin^2 x - 2$

Example

Solve $2\cos^2 x - 5\cos x - 3 = 0$ for $0 \leqslant x \leqslant 2\pi$.

$2\cos^2 x - 5\cos x - 3 = 0$

$(2\cos x + 1)(\cos x - 3) = 0$

$2\cos x + 1 = 0$ or $\cos x - 3 = 0$

Use quadratic factorising.
If it helps you, then replace $\cos x$ by C:
$2C^2 - 5C - 3 = (2C + 1)(C - 3)$

$\cos x = -\frac{1}{2}$ or $\cos x = 3$

The case $\cos x = 3$ has no solutions.

The values of $\cos x$ range from −1 to 1 so there is no angle x for which $\cos x$ has value 3.

The case $\cos x = -\frac{1}{2}$ leads to:

$\cos x$ is negative so you can use the quadrant diagrams at the top of page 38 to find x.

x is in the 2nd or 3rd quadrants

and the 1st quadrant angle is $\frac{\pi}{3}$

So $x = \pi - \frac{\pi}{3}$ or $x = \pi + \frac{\pi}{3}$

So $x = \frac{2\pi}{3}$ or $x = \frac{4\pi}{3}$

6 Match two factors on the right with each quadratic expression on the left.
Some factors will match more than one quadratic expression.

- (A) $\cos^2 x - \cos x - 2$
- (B) $2\cos^2 x + 3\cos x - 9$
- (C) $2\cos^2 x + \cos x$
- (D) $\cos^2 x - 1$
- (E) $2\cos^2 x - 5\cos x + 2$
- (F) $\cos^2 x + 6\cos x + 9$
- (G) $4\cos^2 x - 1$
- (H) $\cos^2 x - \cos x$

- (1) $2\cos x - 1$
- (2) $\cos x + 3$
- (3) $\cos x - 1$
- (4) $\cos x + 1$
- (5) $\cos x$
- (6) $\cos x - 2$
- (7) $2\cos x + 1$
- (8) $2\cos x - 3$

7 Solve for θ where $0 \leqslant \theta \leqslant 360$:

a $2\sin^2 \theta° - \sin \theta° = 0$ **b** $\cos^2 \theta° + \cos \theta° = 0$

c $\tan^2 \theta° - \tan \theta° = 0$ **d** $2\cos^2 \theta° + \cos \theta° = 0$

Hint Look for the common factor.

8 Solve for x where $0 \leqslant x \leqslant 2\pi$:

a $\sin^2 x + \sin x = 0$ **b** $\cos x - 2\cos^2 x = 0$

c $\tan x + \tan^2 x = 0$ **d** $\sin x + 2\sin^2 x = 0$

9 Solve for θ where $0 \leqslant \theta \leqslant 360$:

a $2\cos^2\theta° + 3\cos\theta° + 1 = 0$

b $2\sin^2\theta° - \sin\theta° + 1 = 0$

c $4\sin^2\theta° - 1 = 0$

d $\tan^2\theta° - 2\tan\theta° + 1 = 0$

e $\cos^2\theta° + 2\cos\theta° + 1 = 0$

f $2\cos^2\theta° + 7\cos\theta° + 3 = 0$

g $3\sin^2\theta° + \sin\theta° - 4 = 0$

h $2\sin^2\theta° - \sin\theta° - 1 = 0$

i $\tan^2\theta° - 1 = 0$

j $2\cos^2\theta° + \cos\theta° - 1 = 0$

10 Solve for x where $0 \leqslant x \leqslant 2\pi$:

a $\tan^2 x + 2\tan x + 1 = 0$

b $2\cos^2 x - \cos x - 1 = 0$

c $\sin^2 x - 1 = 0$

d $4\sin^2 x - 4\sin x - 3 = 0$

e $4\cos^2 x + 4\cos x + 1 = 0$

f $2\sin^2 x + 3\sin x + 1 = 0$

g $4\sin^2 x - 4\sin x + 1 = 0$

h $4\cos^2 x - 8\cos x + 3 = 0$

11 Solve for θ where $0 \leqslant \theta \leqslant 360$ giving your answers correct to 1 decimal place:

a $3\sin^2\theta° - 5\sin\theta° - 2 = 0$

b $4\cos^2\theta° + 3\cos\theta° - 1 = 0$

c $\tan^2\theta° - \tan\theta° - 6 = 0$

d $15\sin^2\theta° + 7\sin\theta° - 2 = 0$

12 Solve for x where $0 \leqslant x \leqslant 2\pi$ giving your answers correct to 3 significant figures:

a $4\cos^2 x + 15\cos x + 9 = 0$

b $9\sin^2 x + 3\sin x - 2 = 0$

c $15\cos^2 x + 2\cos x - 1 = 0$

d $2\tan^2 x + 3\tan x - 20 = 0$

Exercise 8.3 Using identities to solve trig equations

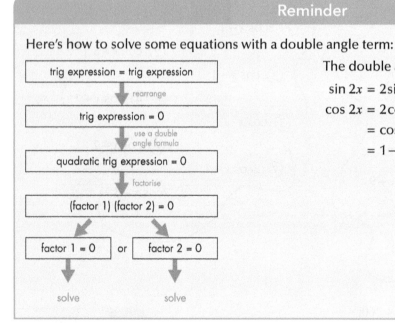

Reminder

Here's how to solve some equations with a double angle term:

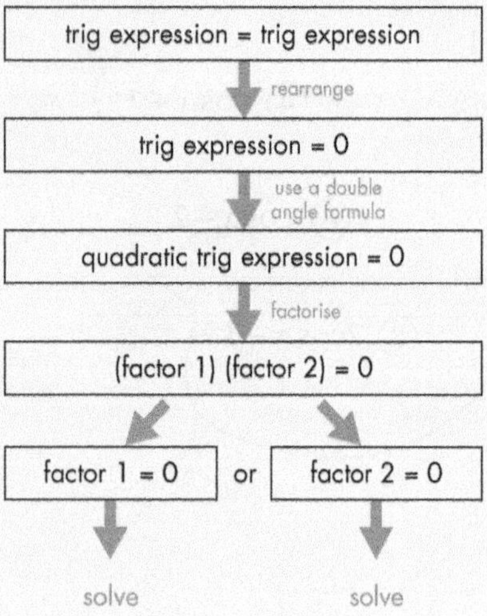

The double angle formulae are:

$\sin 2x = 2\sin x\cos x$

$\cos 2x = 2\cos^2 x - 1$

$= \cos^2 x - \sin^2 x$

$= 1 - 2\sin^2 x$

1 Find values for $\sin x°$ and/or $\cos x°$.

a $\sin 2x° = 0$

b $\sin 2x° - \sin x° = 0$

c $\cos x° + \sin 2x° = 0$

d $\sin 2x° - \cos x° = 0$

e $\sin 2x° + 2\sin x° = 0$

f $\cos x° - 2\sin 2x° = 0$

2 Solve:

a $\sin 2x° - \sin x° = 0$ for $0 \leq x \leq 360$

b $\cos x + \sin 2x = 0$ for $0 \leq x \leq 2\pi$

c $\sin 2x° - \cos x° = 0$ for $0 \leq x \leq 360$

d $\sin 2x + 2\sin x = 0$ for $0 \leq x < 2\pi$

e $2\cos x° - 2\sin 2x° = 0$ for $0 \leq x < 360$

f $2\sin x° + \sin 2x° = 0$ for $0 \leq x < 360$

g $\sqrt{2}\sin 2x + 2\cos x = 0$ for $0 \leq x \leq 2\pi$

h $\sqrt{3}\sin 2x + 3\sin x = 0$ for $\pi \leq x \leq 2\pi$

3 In each case choose the appropriate substitution for $\cos 2x$ and then rearrange into 'normal' quadratic equation order.

a $\cos 2x - \cos x = 0$

b $\cos 2x + \sin x = 0$

c $\sin x - \cos 2x = 0$

d $2 + \cos 2x - \cos x = 0$

e $\cos 2x + 1 = \cos x$

f $\cos 2x = \sin x + 1$

g $3 - 2\cos 2x = 3\cos x + 4$

h $3\cos 2x - 2\sin x + 1 = 0$

4 Solve:

a $\cos 2x - \cos x = 0$ for $0 \leq x \leq 2\pi$

b $\cos 2x° + \sin x° = 0$ for $0 \leq x \leq 360$

c $\cos 2x° - \sin x° = 0$ for $0 \leq x \leq 360$

d $\cos 2x + \cos x + 1 = 0$ for $0 \leq x \leq 2\pi$

e $\cos 2x° - \sin x° - 1 = 0$ for $0 \leq x \leq 360$

f $3\sin x + 2 = \cos 2x$ for $0 \leq x \leq 2\pi$

Exercise 8.4 Applications including the wave function

1 Solve algebraically for $0 \leq x < 360$:

a $\sqrt{7}\cos(x - 35)° = 2$

b $3\sin(x - 15{\cdot}3)° = 1$

c $5\cos(x + 9{\cdot}2)° = -3$

d $\sqrt{3}\sin(x + 24{\cdot}6)° = -0{\cdot}5$

2 $f(x) = 3\cos x° + \sin x°$

a Express $f(x)$ in the form $k\cos(x - \alpha)°$ where $k > 0$ and $0 \leq \alpha < 360$.

b Hence solve algebraically $f(x) = -0{\cdot}5$ for $0 \leq x < 360$.

3 $g(x) = \sqrt{3}\sin x° - \cos x°$

a Express $g(x)$ in the form $k\sin(x - \alpha)°$ where $k > 0$ and $0 \leq \alpha < 360$.

b Hence solve algebraically $g(x) = 0{\cdot}8$ for $0 \leq x < 360$.

4 Solve $4\sin x° - 3\cos x° = 2{\cdot}3$ for $0 \leq x \leq 360$.

5 In each case give the maximum and minimum value of f and the corresponding values of x, where $0 \leq x \leq 360$.

a $f(x) = 2\sin(x + 20)°$

b $f(x) = 3\cos(x - 30)°$

c $f(x) = \sqrt{2}\cos(x + 10)°$

d $f(x) = 2\sqrt{3}\sin(x + 100)°$

e $f(x) = 10\sin(x - 100)°$

f $f(x) = 2{\cdot}6\cos(x + 120)°$

6 Find the maximum and minimum value of these expressions and the corresponding values of θ for $0 \leq \theta \leq 360$.

a $7\sin\theta° + 24\cos\theta°$

b $12\cos\theta° - 5\sin\theta°$

c $3\sin\theta° - 4\cos\theta°$

d $5\sin\theta° + 12\cos\theta°$

e $\cos\theta° - 3\sin\theta°$

f $3\cos\theta° + 2\sin\theta°$

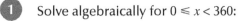

9 Differentiating functions

Exercise 9.1 Differentiating: the power rule

1 Each of these straight line graphs has equation $y = f(x)$. In each case find $f'(x)$.

> **Hint** $f'(x)$ gives the gradient.

a

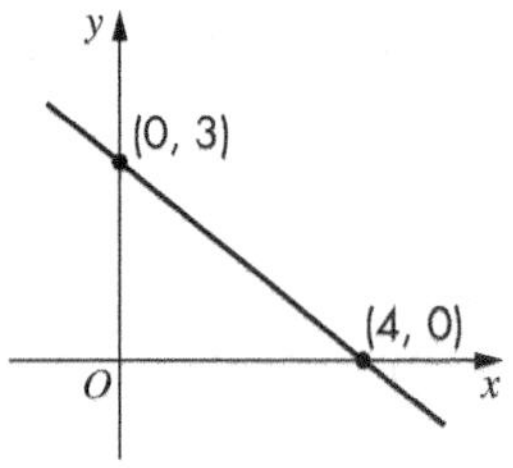

b

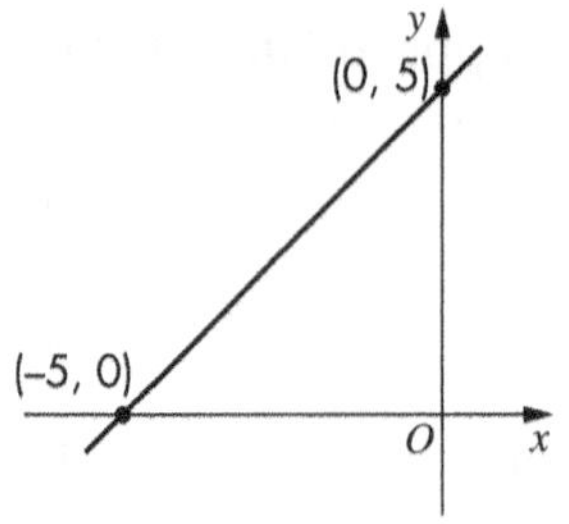

c

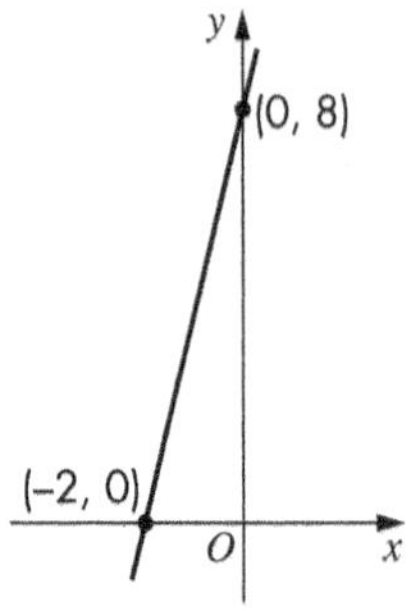

d

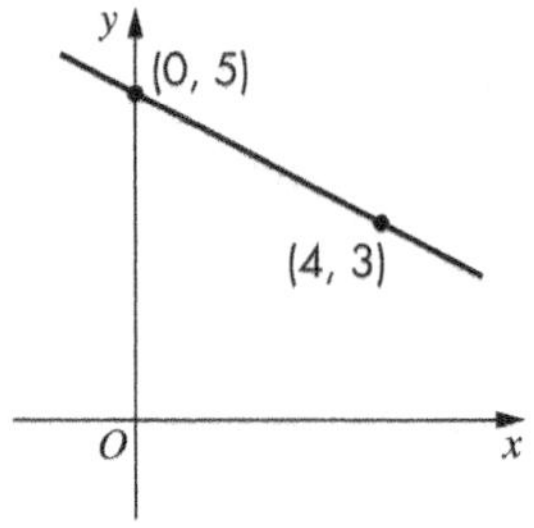

e

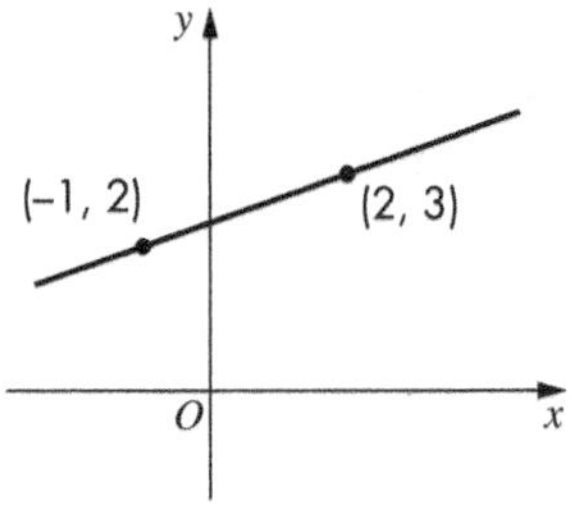

f

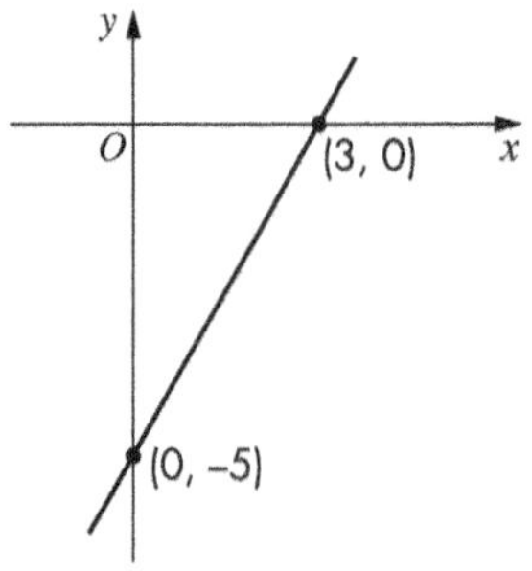

g

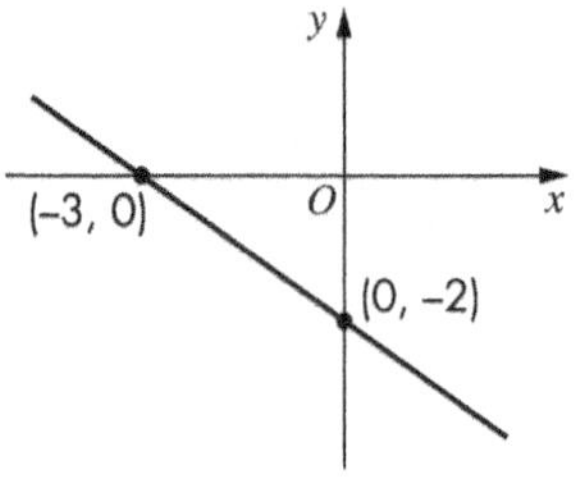

h

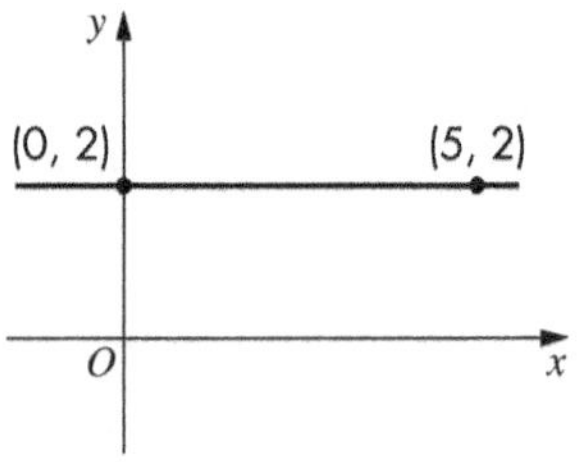

2 Find $f'(x)$.

a $f(x) = x^5$ **b** $f(x) = 3x^5$ **c** $f(x) = 2x^{-1}$

d $f(x) = -x^2$ **e** $f(x) = -3x^4$ **f** $f(x) = 5x$

g $f(x) = -4x^{-3}$ **h** $f(x) = 2$ **i** $f(x) = x^{-2}$

j $f(x) = -x^{-1}$ **k** $f(x) = 10x^2$ **l** $f(x) = \frac{1}{2}x^6$

m $f(x) = \frac{1}{3}x^2$ **n** $f(x) = \frac{1}{2}x^{-1}$ **o** $f(x) = -\frac{1}{4}x^{-8}$

> **Reminder**
>
> $f(x) = x^n$
>
> $\Rightarrow f'(x) = nx^{n-1}$
>
> $f(x) = ax^n$
>
> $\Rightarrow f'(x) = nax^{n-1}$

3 Find $\frac{dy}{dx}$.

a $y = x^7$ **b** $y = 2x^4$ **c** $y = x^{-2}$ **d** $y = -3x^3$

e $y = -2x^3$ **f** $y = 2x$ **g** $y = -2x^{-4}$ **h** $y = 1$

i $y = \frac{3}{2}x^{-2}$ **j** $y = -\frac{1}{2}x^{-6}$ **k** $y = \frac{3}{4}x^4$ **l** $y = \frac{2}{3}x^6$

m $y = \frac{2}{5}x^{-5}$ **n** $y = -\frac{1}{7}x^{-14}$ **o** $y = -\frac{5}{4}x^{-8}$ **p** $y = -\frac{5}{3}x^{12}$

4 Differentiate with respect to x:

a $x^2 - x$ **b** $2x^2 - 2x + 3$ **c** $x^3 - 3x^2 + 3x - 1$

d $-x^4 + 5x - 1$ **e** $\frac{4}{3}x^3 - \frac{1}{2}x^2 + x - 1$ **f** $\frac{1}{3}x^3 + \frac{1}{4}x^2 - x + 1$

g $x^{-2} - x^{-1}$ **h** $6x^{-3} + \frac{1}{2}x^{-2} - 5$ **i** $\frac{3}{2} - \frac{1}{2}x^{-1} + \frac{2}{3}x^{-2}$

5 Match each function with its derived function.

- Ⓐ $f(x) = x^2 - \frac{1}{2}x + 1$
- Ⓑ $f(x) = \frac{1}{2}x^2 - x + 1$
- Ⓒ $f(x) = x - \frac{1}{2}x^2 - 1$
- Ⓓ $f(x) = 1 - x^2 - \frac{1}{2}x$
- Ⓔ $f(x) = 1 - \frac{1}{2}x^2 - x$
- Ⓕ $f(x) = 1 + \frac{1}{2}x - \frac{1}{2}x^2$
- Ⓖ $f(x) = 1 + x + \frac{1}{2}x^2$
- Ⓗ $f(x) = 1 - \frac{1}{2}x^2 - \frac{1}{2}x$
- Ⓘ $f(x) = \frac{1}{2}x^2 - \frac{1}{2}x - 1$

- ① $f'(x) = -2x - \frac{1}{2}$
- ② $f'(x) = \frac{1}{2} - x$
- ③ $f'(x) = x - 1$
- ④ $f'(x) = 1 + x$
- ⑤ $f'(x) = 1 - x$
- ⑥ $f'(x) = -x - \frac{1}{2}$
- ⑦ $f'(x) = 2x - \frac{1}{2}$
- ⑧ $f'(x) = x - \frac{1}{2}$
- ⑨ $f'(x) = -x - 1$

6 Find the derivatives of:

a $x^{\frac{1}{2}}$ **b** $x^{-\frac{1}{2}}$ **c** $x^{\frac{3}{2}}$ **d** $2x^{\frac{3}{2}}$

e $\frac{1}{2}x^{\frac{1}{2}}$ **f** $4x^{-\frac{1}{2}}$ **g** $-x^{-\frac{3}{2}}$ **h** $x^{\frac{1}{3}}$

i $x^{-\frac{1}{3}}$ **j** $6x^{\frac{2}{3}}$ **k** $-3x^{-\frac{2}{3}}$ **l** $-\frac{1}{2}x^{\frac{2}{3}}$

7 Differentiate:

a $f(x) = x^{\frac{1}{2}} + x^{\frac{3}{2}}$ **b** $y = -2x^{\frac{1}{2}} - x^{-\frac{1}{2}}$ **c** $y = 5x^{-\frac{1}{2}} + 2x^{-1}$

d $f(x) = 3x^{-2} + \frac{1}{2}x^{-1}$ **e** $y = \frac{2}{3}x^{\frac{1}{2}} - 2x^{-\frac{1}{2}}$ **f** $f(x) = x^{-3} - x^{-2} - x^{-1}$

g $f(x) = \frac{1}{3}x^{\frac{3}{2}} - \frac{2}{3}x^{-\frac{1}{2}}$ **h** $f(x) = -x^{-\frac{3}{2}} - 3x^{-\frac{1}{2}}$ **i** $f(x) = 6x^{\frac{1}{3}} - 3x^{\frac{4}{3}}$

Exercise 9.2 Differentiation requiring simplification first

1 Differentiate the following by first multiplying out and then simplifying.

a $(x - 2)(x + 4)$ **b** $4x(2x - 5)$ **c** $(3x - 1)^2$

d $(3x - 4)(x + 2)$ **e** $x(x - 6)^2$ **f** $(x - 1)(x + 1)(x - 2)$

g $(x + 1)(x^3 - 3)$ **h** $(x^2 - 1)(x^2 + 1)$ **i** $(2x - 1)(x + 1)(x - 2)$

j $3x^2(x - 5)(x + 2)$ **k** $x^5(x^3 - 1)(x^3 + 1)$ **l** $(3x - 1)(2x + 1)^2$

Hint Remember these rules of indices: $x = x^1$ $\sqrt{x} = x^{\frac{1}{2}}$ $\frac{1}{x^n} = x^{-n}$

2 Find $\frac{dy}{dx}$ where:

a $y = \frac{2}{x^4}$ **b** $y = -\frac{3}{x^2}$ **c** $y = \frac{1}{x^5}$ **d** $y = \frac{5}{x}$

e $y = -\frac{4}{x}$ **f** $y = -\frac{1}{x^3}$ **g** $y = \frac{2}{\sqrt{x}}$ **h** $y = -\frac{1}{\sqrt{x}}$

3 Differentiate with respect to x:

a $\frac{1}{2x}$ **b** $\frac{2}{3x}$ **c** $-\frac{1}{2\sqrt{x}}$ **d** $\frac{1}{3x^2}$

e $-\frac{2}{5x^3}$ **f** $\frac{3}{2\sqrt{x}}$ **g** $\frac{5}{4\sqrt{x}}$ **h** $-\frac{4}{3\sqrt{x}}$

Hint

$$\frac{3}{4x} = \frac{3}{4} \times \frac{1}{x} = \frac{3}{4} \times \frac{x^{-1}}{1} = \frac{3x^{-1}}{4}$$

4 Find $\frac{dy}{dx}$ where:

a $y = \sqrt{x}$ **b** $y = x^2\sqrt{x}$ **c** $y = \left(\sqrt{x}\right)^3$

d $y = \frac{x}{\sqrt{x}}$ **e** $y = \frac{\sqrt{x}}{x}$ **f** $y = x^3\sqrt{x}$

g $y = \frac{1}{x\sqrt{x}}$ **h** $y = \frac{x^2}{\sqrt{x}}$

Reminder

$$a^m \times a^n = a^{m+n}$$

$$\frac{a^m}{a^n} = a^{m-n}$$

$$\left(a^m\right)^n = a^{mn}$$

5 Find $\frac{dy}{dx}$ where:

a $y = \frac{4}{x^3} + \sqrt{x}$ **b** $y = -\frac{3}{x} - x\sqrt{x}$ **c** $y = \frac{1}{x} - \frac{x}{\sqrt{x}}$

d $y = -\frac{5}{x^3} - x^2\sqrt{x}$ **e** $y = x\sqrt{x} + \frac{8}{x}$ **f** $y = \frac{\sqrt{x}}{x} - \frac{3}{x^2}$

g $y = x\sqrt{x} - \frac{1}{x\sqrt{x}}$ **h** $y = \frac{2x^2}{\sqrt{x}} + \frac{\sqrt{x}}{x^2}$ **i** $y = \frac{1}{\sqrt[3]{x}} + \sqrt[3]{x}$

Example

Find $f'(x)$ where $f(x) = \frac{(x-1)^2}{\sqrt{x}}$

$$f(x) = \frac{(x-1)(x-1)}{x^{\frac{1}{2}}} = \frac{x^2 - 2x + 1}{x^{\frac{1}{2}}}$$

Remove the brackets.

Rewrite $\sqrt{x}$ as $x^{\frac{1}{2}}$

$$= \frac{x^2}{x^{\frac{1}{2}}} - \frac{2x^1}{x^{\frac{1}{2}}} + \frac{1}{x^{\frac{1}{2}}}$$

Split the fraction into separate fractions.

$$= x^{2-\frac{1}{2}} - 2x^{1-\frac{1}{2}} + x^{-\frac{1}{2}}$$

Use the laws of indices.

$$= x^{\frac{3}{2}} - 2x^{\frac{1}{2}} + x^{-\frac{1}{2}}$$

The expression can now be differentiated.

$$f'(x) = \frac{3}{2}x^{\frac{3}{2}-1} - \frac{1}{2} \times 2x^{\frac{1}{2}-1} - \frac{1}{2}x^{-\frac{1}{2}-1}$$

Use the power rule: $f(x) = ax^n \Rightarrow f'(x) = nax^{n-1}$

$$= \frac{3}{2}x^{\frac{1}{2}} - x^{-\frac{1}{2}} - \frac{1}{2}x^{-\frac{3}{2}}$$

$$= \frac{3\sqrt{x}}{2} - \frac{1}{\sqrt{x}} - \frac{1}{2\left(\sqrt{x}\right)^3}$$

Rewriting using root signs can be useful when you evaluate $f'(x)$.

6 Differentiate:

a $y = \frac{1}{x}$ **b** $f(x) = \frac{3}{x^{\frac{1}{2}}}$ **c** $f(x) = \frac{2}{\sqrt{x}}$

d $y = \frac{1}{x^2} - \sqrt{x}$ **e** $y = \frac{3x}{\sqrt{x}} - \frac{2}{x}$ **f** $f(x) = \frac{2x+1}{\sqrt{x}}$

g $y = \frac{1-x^2}{\sqrt{x}}$ **h** $f(x) = \frac{3-2x}{x^2}$ **i** $f(x) = \frac{1}{\sqrt{x}} - 2\sqrt{x}$

j $f(x) = \frac{x+1}{\sqrt{x}}$ **k** $y = \frac{3\sqrt{x}+x}{\sqrt{x}}$ **l** $f(x) = \frac{\sqrt{x}-x}{x^{\frac{3}{2}}}$

m $y = \frac{1+2x^2}{\sqrt{x}}$ **n** $f(x) = \frac{x^3 - 2\sqrt{x}}{x}$ **o** $y = \frac{\sqrt{x}-x^2}{x}$

p $f(x) = \frac{x+1-\sqrt{x}}{3\sqrt{x}}$ **q** $y = \frac{(1-x)^2}{\sqrt{x}}$ **r** $f(x) = \frac{(1+\sqrt{x})^2}{\sqrt{x}}$

Exercise 9.3 Applications to gradients of tangents to curves

1 For each curve find the gradient of the tangents at the points A, B and C.

a

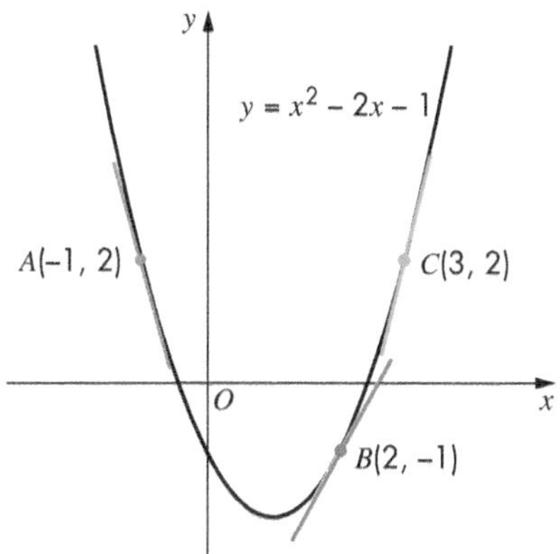

b

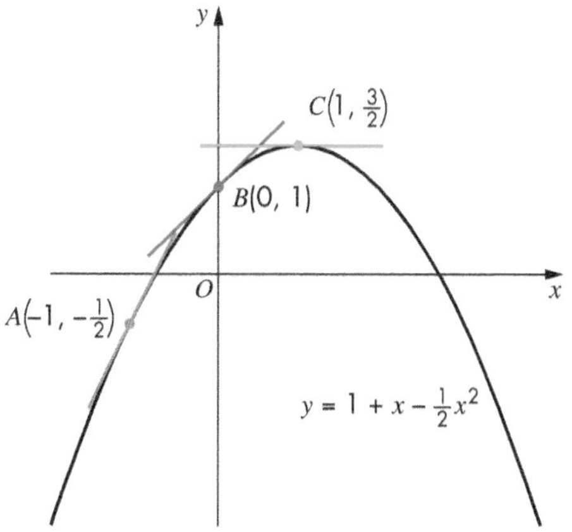

c

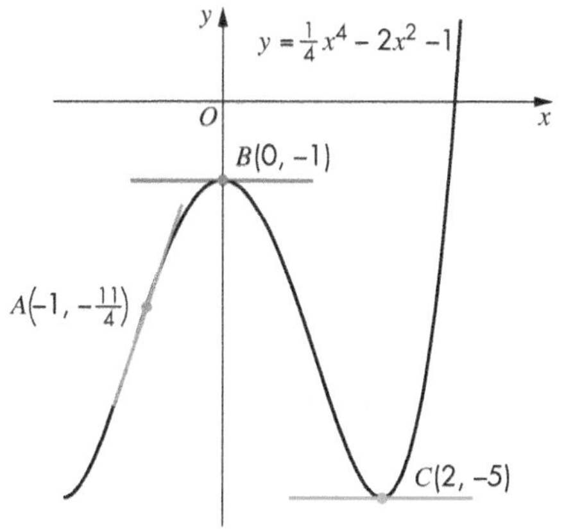

d

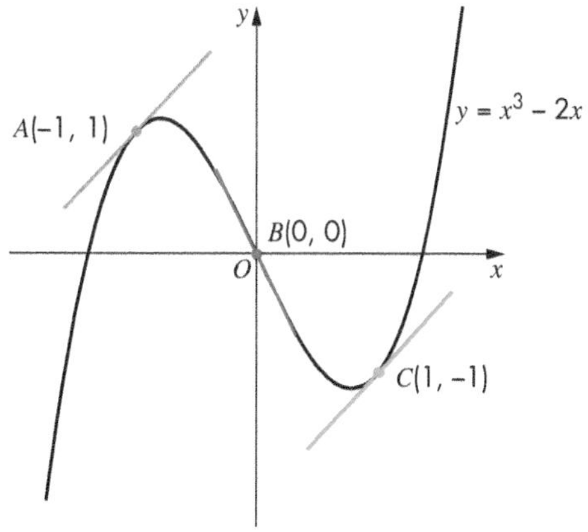

2 **a** $f(x) = x^3 - 2x^2 - 3x + 4$. Find the value of $f'(-1)$.

b Find the value of $g'(2)$ where $g(x) = (2x - 1)(x + 3)$.

c For $y = 2x - \frac{1}{x}$ find the value of $\frac{dy}{dx}$ where $x = \frac{1}{2}$

d Find $f'(4)$ for the function $f(x) = 2\sqrt{x}$.

3 For the graphs shown in each of these diagrams, calculate the gradient of the tangent to the graph at each of the indicated points A, B and C.

a

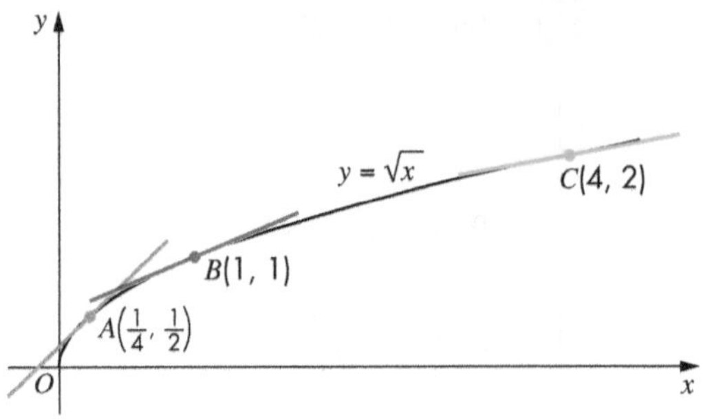

b

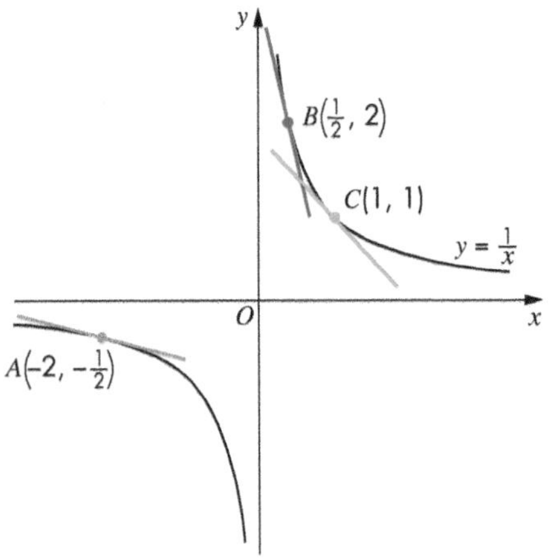

c

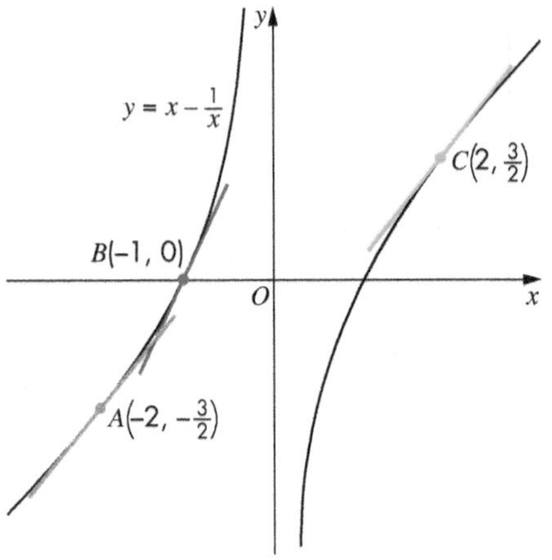

4 Find:

a $f'(4)$ where $f(x) = 5\sqrt{x} - x$

b $f'(16)$ where $f(x) = \dfrac{\sqrt{x} - 4x^2}{\sqrt{x}}$

c $f'(1)$ where $f(x) = \dfrac{x - \sqrt{x}}{x}$

d $f'(9)$ where $f(x) = \dfrac{x^2 - x}{\sqrt{x}}$

e $f'(1)$ where $f(x) = \dfrac{2 + x}{\sqrt{x}}$

f $f'(4)$ where $f(x) = \dfrac{7\sqrt{x} - 2x}{\left(\sqrt{x}\right)^3}$

g $f'(36)$ where $f(x) = \dfrac{2x + 3x^2}{\sqrt{x}}$

h $f'(9)$ where $f(x) = \dfrac{18 - 3x}{2\sqrt{x}}$

5 Find the coordinates of the two points on the curve $y = x^3 - 2x + 3$ where the gradient is 10.

6 There are two points on the curve $y = 2x^3 - x^2$ where the gradient is 4. Find the coordinates of these points.

7 Where on the curve $y = -\dfrac{2}{x}$ is the gradient $\dfrac{2}{9}$?

8 A tangent to the curve $y = \sqrt{x}$ has gradient $\frac{1}{3}$. Find its point of contact with the curve.

9 The diagram shows part of the graph $y = 2x^3 + 3x^2 - 12x$.

Find the coordinates of the two points A and B on the graph where the gradient of the tangent to the graph has value zero.

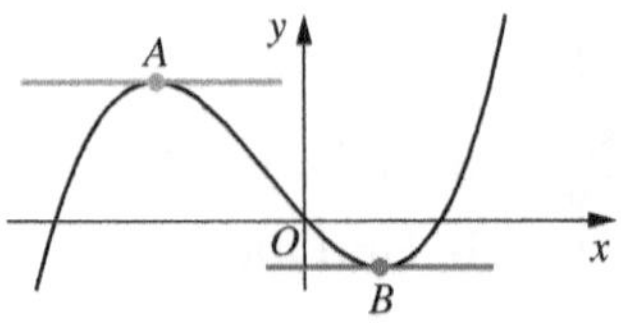

Exercise 9.4 Differentiation: the chain rule

Reminder

The chain rule can be used to extend the scope of the basic power rule:

The power rule: $y = x^n \Rightarrow \dfrac{dy}{dx} = nx^{n-1}$

The extended power rule: $y = \left(f(x)\right)^n \Rightarrow \dfrac{dy}{dx} = n\left(f(x)\right)^{n-1} \times f'(x)$

An example: $y = \sqrt{2x-1} = (2x-1)^{\frac{1}{2}} \Rightarrow \dfrac{dy}{dx} = \dfrac{1}{2}(2x-1)^{\frac{1}{2}-1} \times 2 = (2x-1)^{-\frac{1}{2}} = \dfrac{1}{\sqrt{2x-1}}$

1 Differentiate with respect to x:

a $(2x-1)^4$ **b** $(1-3x)^5$ **c** $(2+x+x^2)^3$

d $(1-x-x^2)^5$ **e** $(6x+3)^3$ **f** $(2x+1)^{-1}$

g $(1-x)^{-2}$ **h** $(x+5)^{-3}$ **i** $3(3x-1)^5$

j $2(4-x)^3$ **k** $-(4x-x^2)^3$ **l** $2(1-x+x^2)^{-1}$

m $-3(1-x^3)^5$ **n** $-2(2x+7)^{-3}$ **o** $\frac{1}{2}(5x-2)^6$

2 Differentiate with respect to the given variable:

a $(t+1)^6$ **b** $(2w+1)^4$ **c** $(5+m)^5$

d $(3p-1)^8$ **e** $(1-y)^{-2}$ **f** $(2-3x)^{-3}$

g $(4-w)^{-2}$ **h** $(1-3k)^5$ **i** $(y^2-1)^4$

j $(r^3-2)^3$ **k** $(t^4-3)^4$ **l** $(1-m^4)^3$

m $(4x+1)^{\frac{1}{2}}$ **n** $(6y-1)^{\frac{1}{3}}$ **o** $(x-2)^{-3}$

p $(1-6x)^{\frac{1}{2}}$ **q** $(1+5f)^{-2}$ **r** $(k^2-2k)^4$

3 The diagram shows part of the graph of the function

$f(x) = \dfrac{1}{1-2x}$

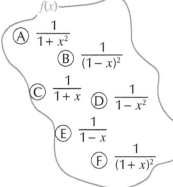

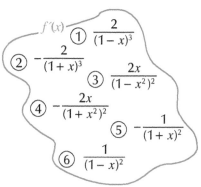

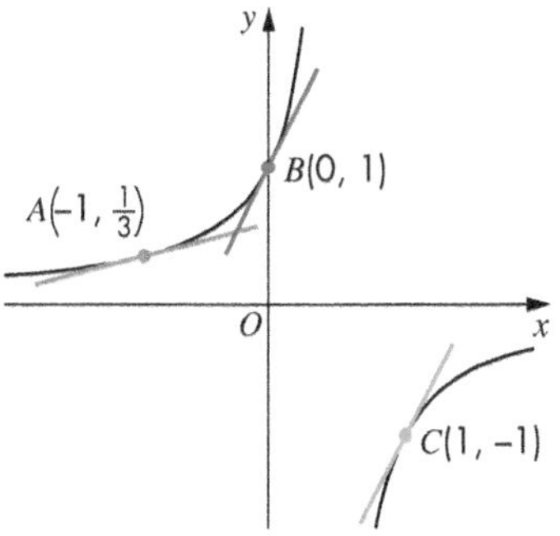

a Rewrite $f(x)$ as a power of $1-2x$.

b Find $f'(x)$ in a form that uses a positive index.

c For each point A, B and C, calculate the exact value of the gradient of the tangent to the curve at that point.

4 Match each function with its derived function.

$f(x)$

- Ⓐ $\dfrac{1}{1+x^2}$
- Ⓑ $\dfrac{1}{(1-x)^2}$
- Ⓒ $\dfrac{1}{1+x}$
- Ⓓ $\dfrac{1}{1-x^2}$
- Ⓔ $\dfrac{1}{1-x}$
- Ⓕ $\dfrac{1}{(1+x)^2}$

$f'(x)$

- ① $\dfrac{2}{(1-x)^3}$
- ② $-\dfrac{2}{(1+x)^3}$
- ③ $\dfrac{2x}{(1-x^2)^2}$
- ④ $-\dfrac{2x}{(1+x^2)^2}$
- ⑤ $-\dfrac{1}{(1+x)^2}$
- ⑥ $\dfrac{1}{(1-x)^2}$

5 Find $f'(x)$ for the following.

a $f(x) = \frac{1}{2x+1}$
b $f(x) = \frac{2}{1-x}$
c $f(x) = -\frac{3}{2x+1}$
d $f(x) = x - \frac{1}{1-x}$
e $f(x) = x + \frac{3}{2x+1}$
f $f(x) = \frac{1}{(2x-1)^2}$
g $f(x) = \frac{2}{3(1-x)}$
h $f(x) = -\frac{3}{4(x+2)}$
i $f(x) = 1 + \frac{2}{5(1-x)}$
j $f(x) = x + \frac{1}{2x-1}$
k $f(x) = x^2 - \frac{1}{1-2x}$
l $f(x) = \frac{1}{1-x^3}$
m $f(x) = \frac{1}{x^2+1}$
n $f(x) = -\frac{1}{1-x^2}$
o $f(x) = x - \frac{2}{3(x^2-1)}$
p $f(x) = \frac{1}{x+1} + \frac{1}{x-1}$
q $f(x) = \frac{1}{(x+1)^2} - x$
r $f(x) = \frac{1}{x} - \frac{1}{x-1}$

6 Part of the graph $y = \frac{1}{\sqrt{1-2x}}$

is shown in the diagram.

Find the gradients of the tangents at the three points A, B and C.

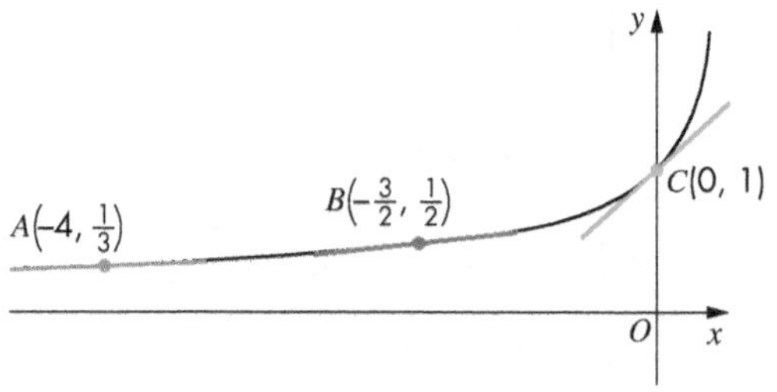

7 The diagram shows part of the graph $y = \frac{1}{1+x} - \frac{1}{1-x}$

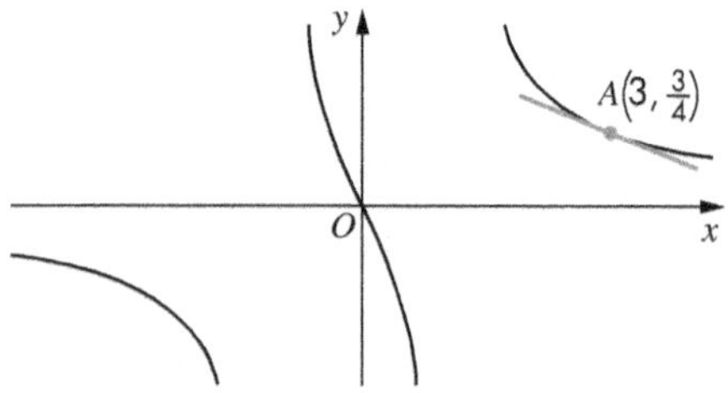

a Find the gradient of the tangent to the graph at the origin.

b Show that the gradient of the tangent to the graph at the point $A\left(3, \frac{3}{4}\right)$ is $-\frac{5}{16}$

Exercise 9.5 Differentiation of some trigonometric functions

1 Differentiate:

a $\sin x$
b $3\sin x$
c $3 - \sin x$
d $5 + \sin x$
e $x + 2\sin x$
f $3x - 4\sin x$
g $\cos x$
h $-2\cos x$
i $4\cos x$
j $7 - \cos x$
k $x + 3\cos x$
l $x^2 - 3\cos x$
m $\sin x - \cos x$
n $\cos x - \sin x$
o $\cos x + \sin x$
p $\sin x + \cos x$
q $x - 3\cos x - 2$
r $2\sin x + 3\cos x$
s $3x^2 + 3\cos x$
t $3\sin x - 2\cos x$
u $-\cos x - x^3$

Reminder

$f(x) = \sin x$

$f'(x) = \cos x$

$f(x) = \cos x$

$f'(x) = -\sin x$

These rules work only if x is measured in radians, not degrees.

2 Find the exact value of:

a $\sin\frac{\pi}{6}$ b $\cos\frac{2\pi}{3}$ c $\cos\pi$

d $\sin\frac{5\pi}{3}$ e $\sin\frac{\pi}{2}$ f $\cos\frac{5\pi}{6}$

g $\sin\frac{\pi}{4}$ h $\sin\frac{7\pi}{4}$ i $\cos\frac{11\pi}{6}$

j $\cos\frac{3\pi}{2}$ k $\sin\frac{3\pi}{4}$ l $\cos\frac{7\pi}{6}$

Hint Use these diagrams along with the quadrant diagrams on page 38:

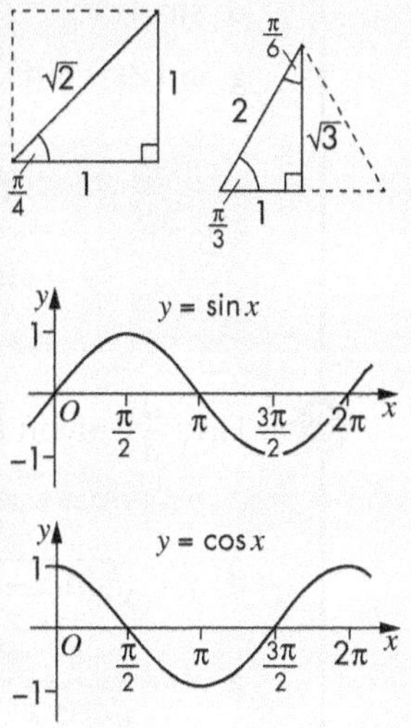

3 Find the exact value of the gradient of the tangent to the graph $y = \sin x$ at the points A, B, C and D as shown in the diagram.

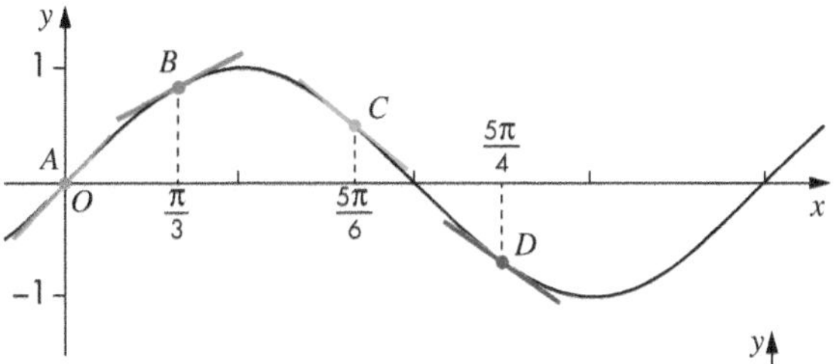

4 Find the exact value of the gradient of each of the four tangents to the graph $y = \cos x$ at the points A, B, C and D as shown in the diagram.

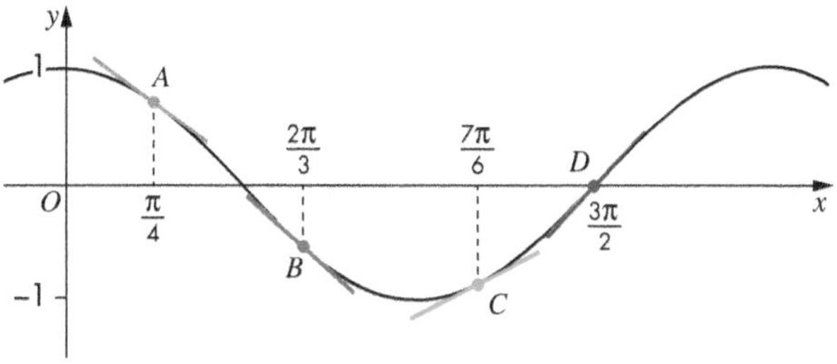

5 $f(x) = \sin x + \cos x$

Find the exact value of the following. Simplify your answer to a single fraction where necessary.

a $f'\left(\frac{\pi}{6}\right)$ b $f'\left(\frac{\pi}{3}\right)$ c $f'\left(\frac{\pi}{4}\right)$

d $f'\left(\frac{3\pi}{2}\right)$ e $f'\left(\frac{5\pi}{6}\right)$ f $f'\left(\frac{5\pi}{3}\right)$

g $f'(2\pi)$ h $f'\left(\frac{5\pi}{4}\right)$ i $f'(\pi)$

Reminder

The chain rule can be used to extend the scope of the differentiation rules for $\sin x$ and $\cos x$:

The rule for $\sin x$: $y = \sin x \Rightarrow \frac{dy}{dx} = \cos x$

The extended rule: $y = \sin(f(x)) \Rightarrow \frac{dy}{dx} = \cos(f(x)) \times f'(x)$

An example: $y = \sin\left(2x - \frac{\pi}{6}\right) \Rightarrow \frac{dy}{dx} = \cos\left(2x - \frac{\pi}{6}\right) \times 2$

6 Differentiate:

a $\sin 2x$ b $\cos 3x$ c $\sin\frac{1}{2}x$ d $2\sin\frac{1}{2}x$

e $-\cos 2x$ f $5\cos 3x$ g $-2\sin\frac{3}{2}x$ h $1 - \frac{1}{2}\cos 2x$

i $1 + \frac{1}{2}\sin 4x$ j $4\sin\frac{x}{2}$ k $2\cos\frac{x}{2}$ l $x - \frac{1}{2}\cos\frac{x}{2}$

Hint $\frac{x}{2} = \frac{1}{2}x$

7 Differentiate:

a $\sin^3 x$ b $\cos^2 x$ c $\frac{1}{2}\sin^2 x$

d $\sin 4x$ e $\sin x - \sin^2 x$ f $\cos x - \cos^2 x$

g $\sin 2x + \sin^2 x$ h $\sin^2 x - \cos^2 x$ i $\sin^2 x + \cos^2 x$

> **Hint** To differentiate $\sin^2 x$ you use the extended power rule:
> $y = \sin^2 x = (\sin x)^2 \Rightarrow \frac{dy}{dx} = 2(\sin x)^1 \times \cos x$

8 Find $\frac{dy}{dx}$ given that:

a $y = (\sin x + x)^3$ b $y = (x^2 - \cos x)^2$ c $y = \sqrt{x^2 + x}$

d $y = \sqrt{\sin x - 3}$ e $y = \sqrt{5 - 3\sin x}$ f $y = \cos^2 x$

g $y = \frac{1}{\sqrt{\cos x}}$ h $y = \frac{1}{\sqrt{\sin x}}$

i $y = \sqrt{2\cos x - 3}$ j $y = \frac{1}{\cos x}$

k $y = \frac{1}{\cos x + \sin x}$ l $y = \frac{1}{\sin^2 x}$

> **Hint** In part **g** rewrite as $y = (\cos x)^{-\frac{1}{2}}$
> In part **j** rewrite as $y = (\cos x)^{-1}$

9 The graphs $y = \sin 2x$ and $y = \sin\frac{1}{2}x$ are shown in the diagram.
Calculate the exact value of the gradient of all the eight tangents shown.

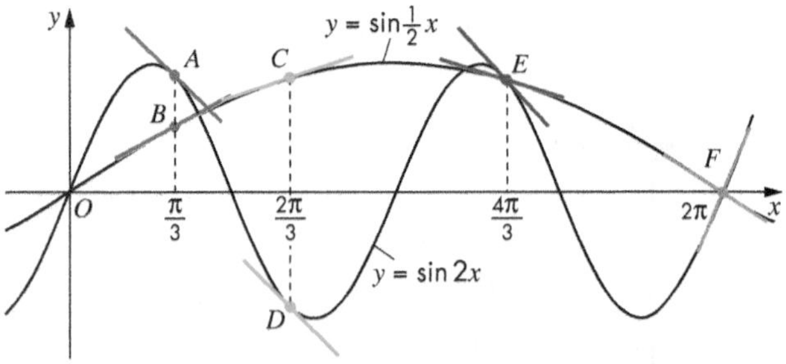

10 Differentiate each of the following in two different ways.

a $2\cos^2 x - 1$ b $1 - 2\sin^2 x$

c $\cos^2 x - \sin^2 x$

> **Hint** The double angle formulae are given on page 42.

11 $f(x) = \sin^2 x$ and $g(x) = \cos^2 x$.

a Find $f'(x)$.

b Find $g'(x)$.

c Hence find $f'(x) + g'(x)$.

d Explain your answer to part **c**.

12 Point A lies on the graph $y = \frac{1}{\sin x}$

a If A has x-coordinate $\frac{\pi}{4}$ find the exact value of its y-coordinate.

b Find an expression for $\frac{dy}{dx}$

c Find the gradient of the tangent to the graph at point A.

d Show that when $x = \frac{4\pi}{3}$, $\frac{dy}{dx} = \frac{2}{3}$

e Find the coordinates of a point on the graph where the gradient of the tangent is 0.

10 Using differentiation to investigate the nature and properties of functions

Exercise 10.1 Determining the equation of a tangent to a curve

Reminder

(a, b) is a point on the line. m is the gradient of the line. The equation of the line is: $y - b = m(x - a)$

1 In each case find the equation of the line through the given point with the given gradient.

Give your answer where possible in the form $ax + by = c$ where a, b, and c are integers.

a point: (1, 2) gradient: –2

b point: (–1, 3) gradient: 3

c point: (0, 1) gradient: 5

d point: (–3, 0) gradient: –1

e point: (2, 1) gradient: $\frac{1}{2}$

f point: $\left(\frac{1}{2}, \frac{1}{2}\right)$ gradient: 2

g point: $\left(-\frac{1}{2}, \frac{3}{2}\right)$ gradient: –1

h point: $\left(1, \frac{5}{3}\right)$ gradient: $-\frac{1}{3}$

i point: $\left(5, -\frac{1}{2}\right)$ gradient: 3

Hint For the x-axis intercept set $y = 0$ in the equation.

For the y-axis intercept set $x = 0$ in the equation.

2 Determine the coordinates of the x- and y-axis intercepts for the following lines.

a $2x - 3y = 6$

b $-x + 3y = 3$

c $5x + y = 10$

d $2x - 3y = 2$

e $-2x - y = 3$

f $6x - 2y = 3$

Example

Find the equation of the tangent to the parabola $y = -\frac{1}{4}x^2 + \frac{1}{2}x + 3$ at the point (–2, 1).

Find the coordinates of the x- and y-axis intercepts of the tangent.

$y = -\frac{1}{4}x^2 + \frac{1}{2}x + 3$

$\frac{dy}{dx} = -\frac{1}{2}x + \frac{1}{2}$ — Differentiate to find the gradient formula.

When $x = -2$, $\frac{dy}{dx} = -\frac{1}{2} \times (-2) + \frac{1}{2} = \frac{3}{2}$ — Use the x-coordinate of the point of contact of the tangent.

point: (–2, 1) gradient: $\frac{3}{2}$ — It's sensible to write down the point and the gradient before you use $y - b = m(x - a)$ so that you don't mix the numbers up.

Equation is: $y - 1 = \frac{3}{2}(x - (-2))$

$\Rightarrow 2y - 2 = 3(x + 2)$ — Double both sides to remove the fraction.

$\Rightarrow 2y - 3x = 8$ — It's useful to tidy up the equation into the form $ax + by = c$.

When $x = 0$, $2y - 0 = 8$ so $y = 4$ — $x = 0$ gives the y-intercept.

When $y = 0$, $0 - 3x = 8$ so $x = -\frac{8}{3}$ — $y = 0$ gives the x-intercept.

Intercepts are (0, 4) and $\left(-\frac{8}{3}, 0\right)$. — You were asked for coordinates.

3 For each of these diagrams:

i find the equation of the tangent at the point P

ii find the coordinates of the axes intercepts A and B.

Check that your answers make sense by comparing them with the diagram.

a

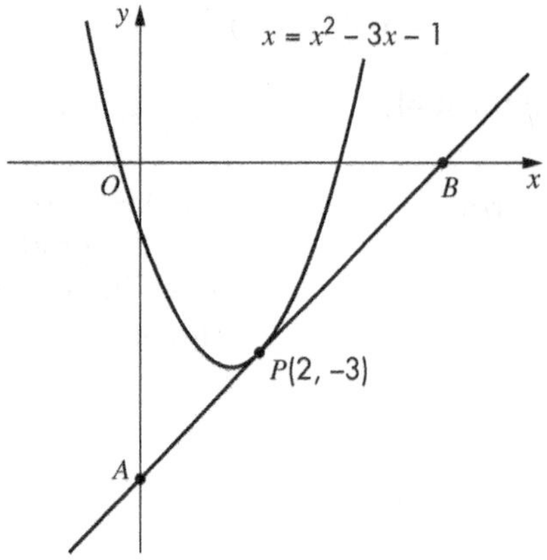

b

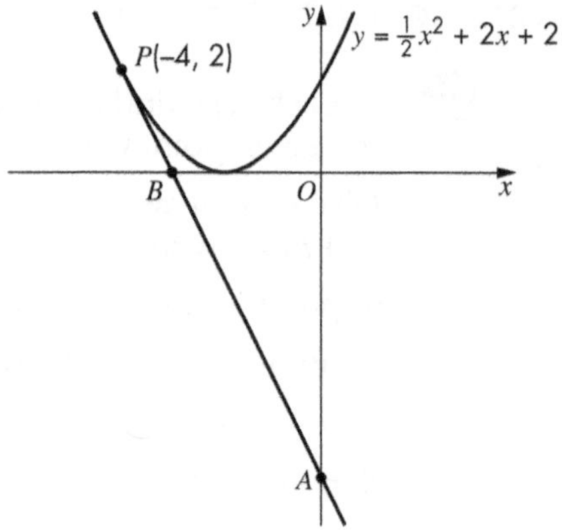

c

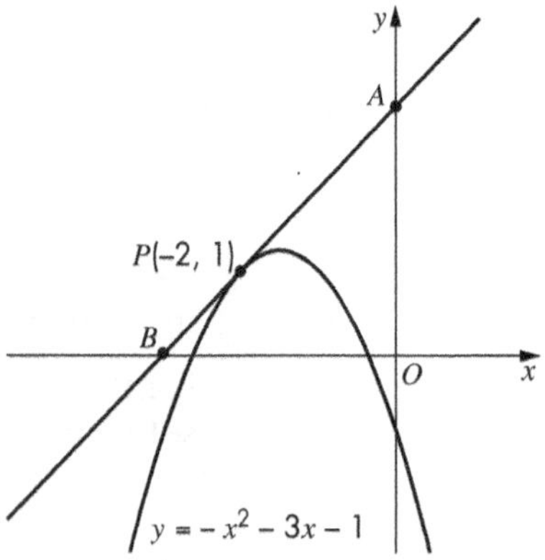

d

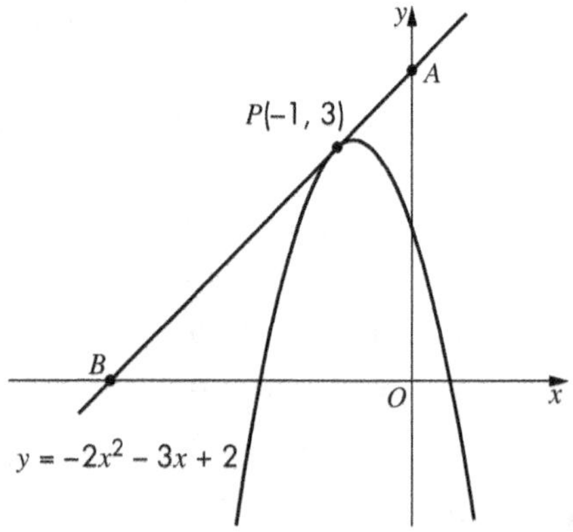

e

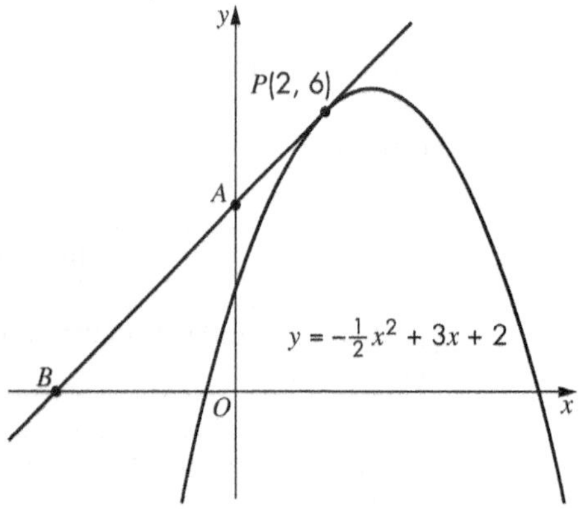

f

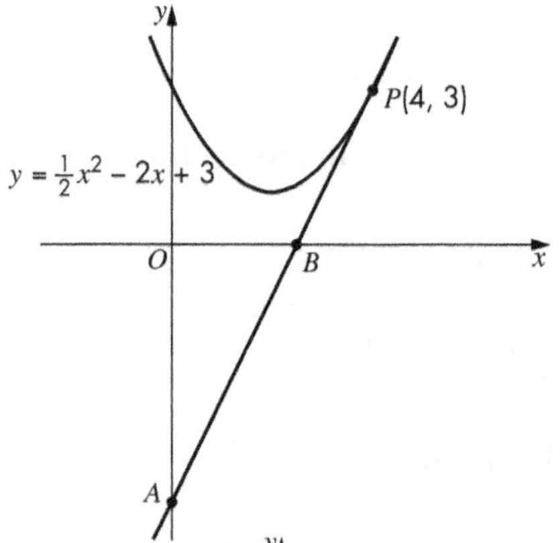

g

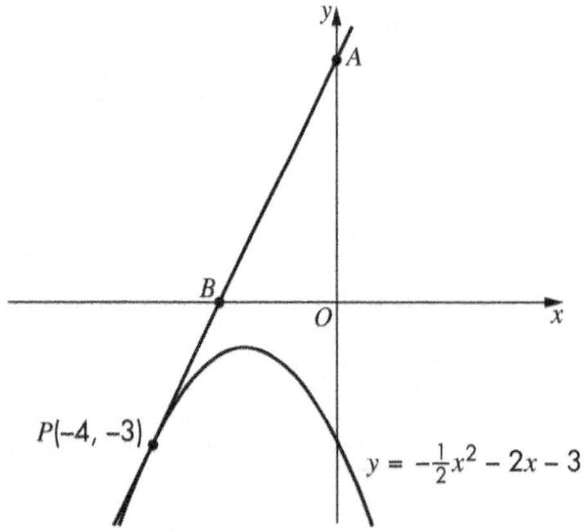

h

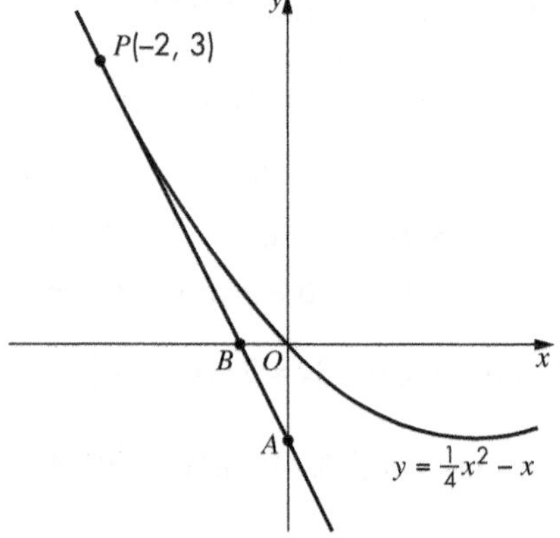

i

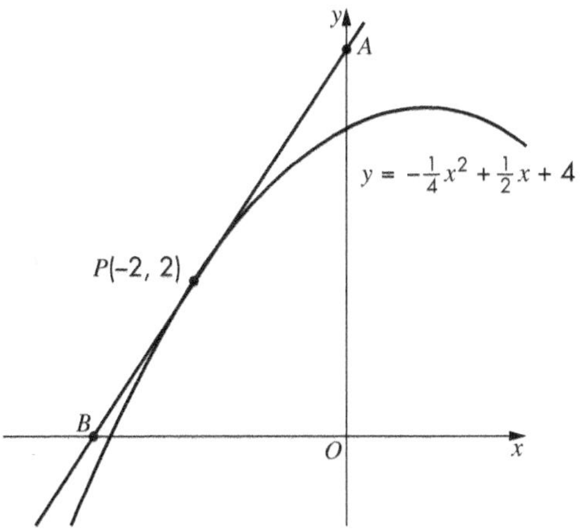

j

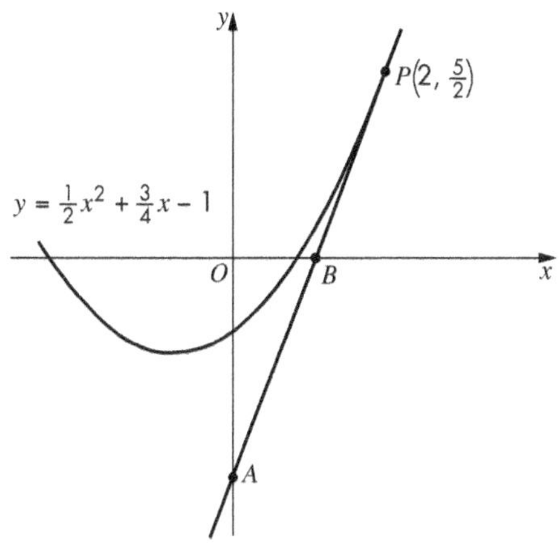

4 In each case find the equation of the tangent at A on the given curve.

a $A(-1, 3)$, $y = 2x^2 - 3x - 2$ **b** $A(2, 6)$, $y = x^3 - 2x^2 + 3x$

c $A(0, 7)$, $y = 2x^3 - 2x + 7$ **d** $A(-3, 1)$, $y = x^3 + 3x^2 - 2x - 5$

5 Find the equation of the tangent to the curve:

a $y = x^2$ at $x = 2$ **b** $y = 4x$ at $x = -1$

c $y = \sqrt{x}$ at $x = 4$ **d** $y = \dfrac{1}{x}$ at $x = 1$

e $y = x^3$ at $x = 2$ **f** $y = 3 - x^2$ at $x = 2$

g $y = x^2 + 5x$ at $x = -1$ **h** $y = x^2 - 2x - 2$ at $x = 3$

i $y = x^3 - 4x + 4$ at $x = -2$ **j** $y = (2x - 1)(x + 2)$ at $x = 1$

k $y = 4x^4$ at $x = \frac{1}{2}$ **l** $y = \dfrac{2}{x^2}$ at $x = -\frac{1}{2}$

Hint For the tangent to $y = f(x)$ at $x = a$:

- point: (a, b) Use $f(a)$ to find b.
- gradient: m Use $f'(a)$ to find m.

Then use $y - b = m(x - a)$

6 On the curve $y = \frac{1}{3}x^3 - 2x + 1$ tangents are drawn at the points $A\left(-1, \frac{8}{3}\right)$, $B(0, 1)$ and $C\left(1, -\frac{2}{3}\right)$.

a Which pair of tangents are parallel?

b Find the equations of the three tangents.

Hint Parallel lines have equal gradients.

7 The diagram shows part of the graph $y = \sin x$.

a Find the equation of the tangent at (0, 0).

b Find the equation of the tangent at (π, 0).

c Find the coordinates of point P, the point of intersection of the two tangents.

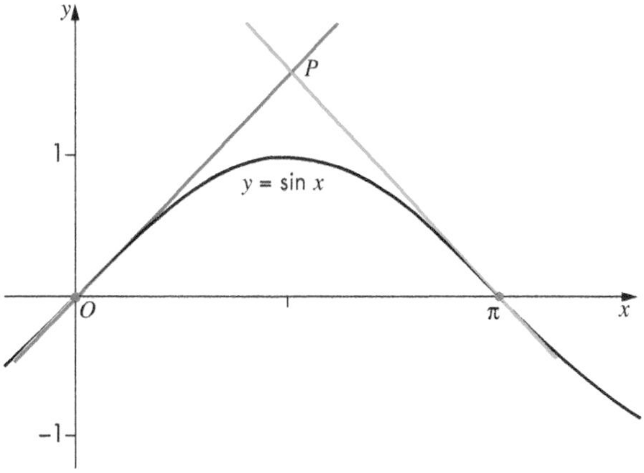

Hint Points of intersection are found by setting the two equations equal to each other and then solving for x.

Exercise 10.2 Investigating decreasing and increasing functions

1 For the given value of x determine whether the graph of the function f is increasing, stationary or decreasing.

a $f(x) = 5x^2 - 2x + 7,\ x = \frac{1}{2}$

b $f(x) = 15x - 3x^3,\ x = 1$

c $f(x) = x^3 - 2x^2 - 6x,\ x = -1$

d $f(x) = 5 - 3x^2 - x^3,\ x = -2$

e $f(x) = -16x^5 - x^2,\ x = \frac{1}{2}$

f $f(x) = 4x^3 - 2x^2 - 8x,\ x = 1$

g $f(x) = 4x^3 - 2x^2 - 8x,\ x = \frac{1}{2}$

h $f(x) = 4x^3 - 2x^2 - 8x,\ x = \frac{3}{2}$

Hint

If $f'(a) < 0$ then f is decreasing at $x = a$.

If $f'(a) = 0$ then f is stationary at $x = a$.

If $f'(a) > 0$ then f is increasing at $x = a$.

Reminder

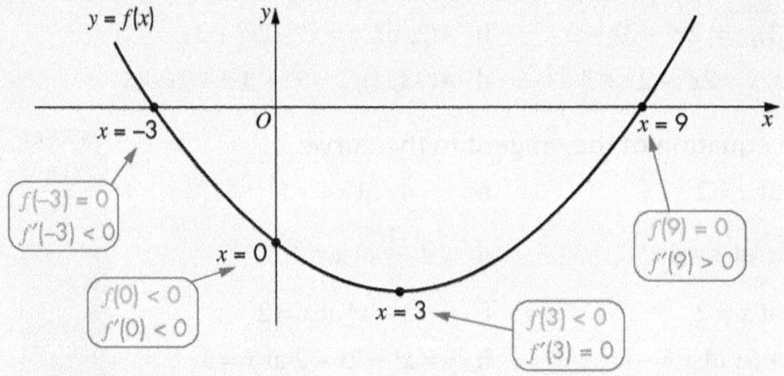

$f(a)$ gives the value of the function f when $x = a$ and is shown by the height of the point on the graph. It is the y-coordinate of the point.

$f'(a)$ gives the value of the gradient when $x = a$ and shows what the slope of the graph (from left to right) is doing at the point:

downhill for $f'(a) < 0$

horizontal for $f'(a) = 0$

uphill for $f'(a) > 0$

2 For each indicated point $x = a$ on the graph $y = f(x)$, choose two true statements from:

$f(a) < 0$ $\quad f(a) = 0$ $\quad f(a) > 0$ $\quad f'(a) < 0$ $\quad f'(a) = 0$ $\quad f'(a) > 0$

a

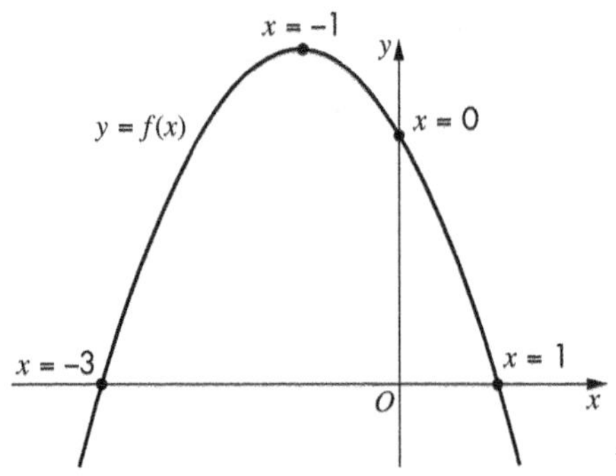

b

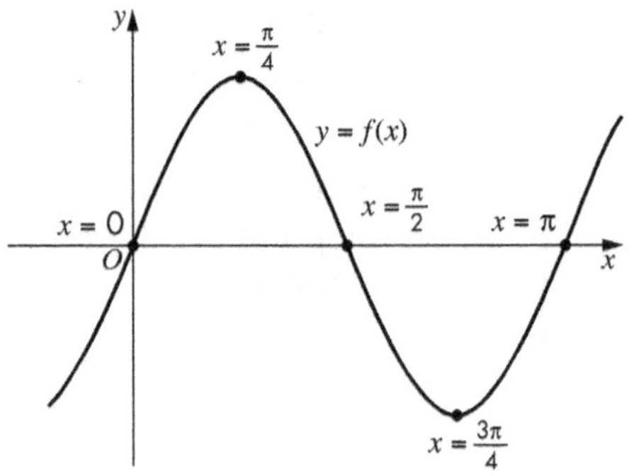

c

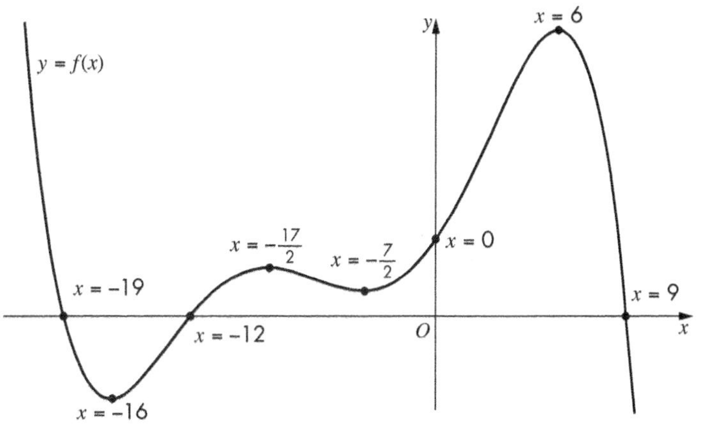

3 For the following quadratic graphs $y = f(x)$, give the values or range of values of x for which:

i f is decreasing **ii** f is stationary **iii** f is increasing

a

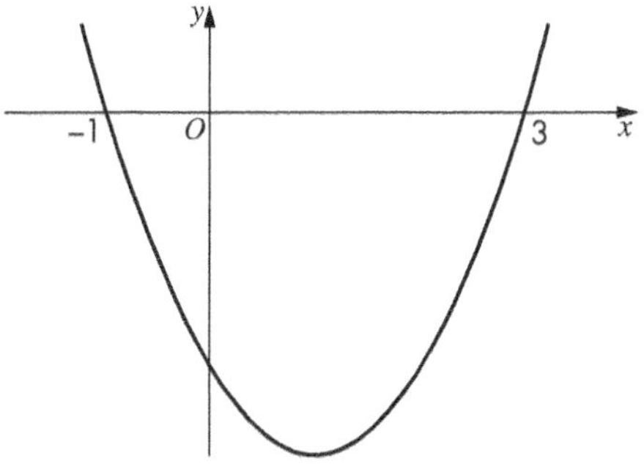

b

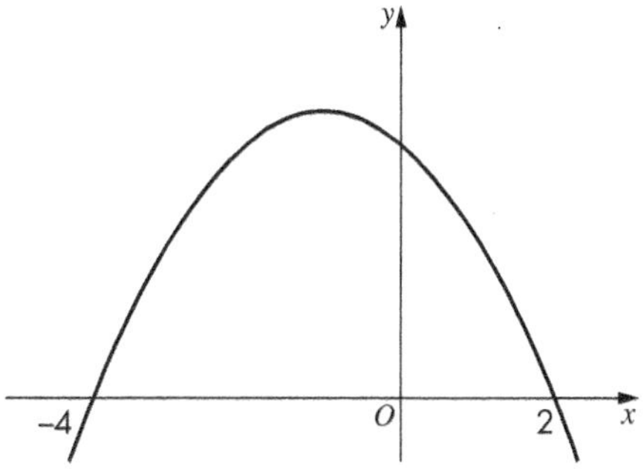

c

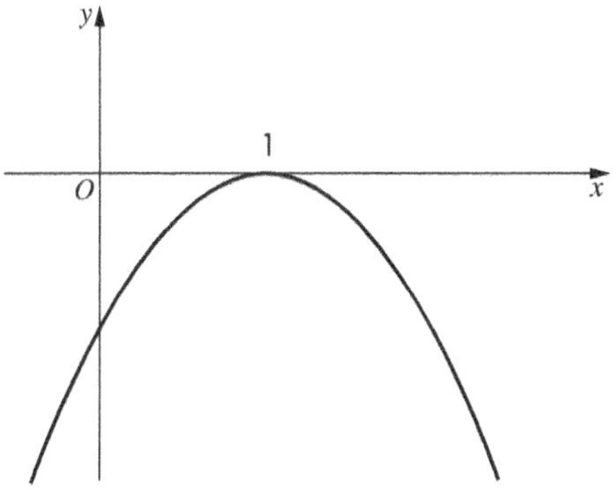

d

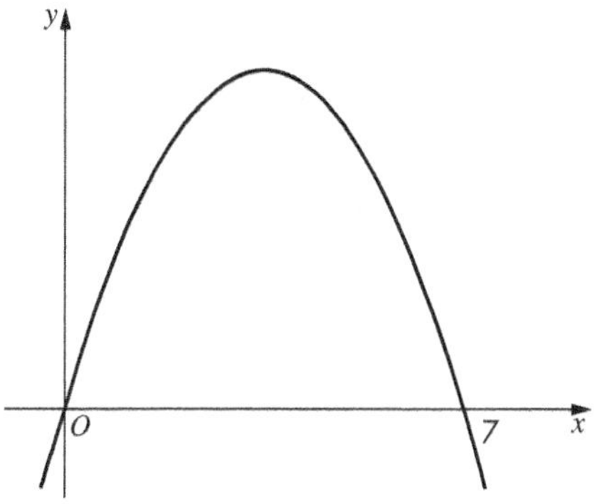

4 Sketch these quadratic graphs to find the values or range of values of x for which the graph is:

i decreasing **ii** stationary **iii** increasing

a $y = -x^2$ **b** $y = x^2 + 2x$ **c** $y = x^2 + 3x - 10$

d $y = x - x^2$ **e** $y = x^2 + 5x + 6$ **f** $y = 3 - 4x - 4x^2$

5 For the given value of x, determine whether the function f is decreasing, stationary or increasing:

a $f(x) = \dfrac{3}{x}$ when $x = 1$

b $f(x) = x - \dfrac{1}{x}$ when $x = 1$

c $f(x) = \dfrac{1}{2x - 1}$ when $x = -\dfrac{1}{2}$

d $f(x) = 2x^2 + \dfrac{1}{2x}$ when $x = \dfrac{1}{2}$

e $f(x) = \sin x + \cos x$ when $x = \dfrac{\pi}{4}$

f $f(x) = \sin\left(2x - \dfrac{\pi}{6}\right)$ when $x = \dfrac{\pi}{3}$

g $f(x) = \frac{1}{2}\cos^2 x$ when $x = \dfrac{\pi}{6}$

h $f(x) = \dfrac{1}{\cos x}$ when $x = 0$

Exercise 10.3 Stationary points and their nature: sketching graphs

Example

Show that $x = 3$ gives a stationary point on the graph $y = f(x)$ where $f(x) = x^4 - 4x^3$.

Find the coordinates and nature of the point.

$f(x) = x^4 - 4x^3 = x^3(x - 4)$

$f'(x) = 4x^3 - 12x^2 = 4x^2(x - 3)$

These factorised forms are easier to use for calculations.

This formula is used to calculate the gradient of the graph.

The calculations:

$f(3) = 3^3(3 - 4) = 27 \times (-1) = -27$ — This calculation gives the y-coordinate of the point.

$f'(3) = 4 \times 3^2(3 - 3) = 4 \times 9 \times 0 = 0$ — This is the gradient of the graph when $x = 3$.

$f'(2) = 4 \times 2^2(2 - 3) = 4 \times 4 \times (-1) = -16 < 0$ (–ve) — This is the gradient just to the left of $x = 3$.

$f'(4) = 4 \times 4^2(4 - 3) = 4 \times 16 \times 1 = 64 > 0$ (+ve) — This is the gradient just to the right of $x = 3$.

Since $f'(3) = 0$ there is a stationary point when $x = 3$. — Essential justification for $x = 3$ giving a stationary point.

The nature table:

x	3^-	3	3^+
$f'(x)$	–	0	+
Shape	\	—	/

For $x = 3^-$ use the $f(2)$ calculation.

For $x = 3^+$ use the $f(4)$ calculation.

–ve from $f(2) = -16 < 0$; +ve from $f(4) = 64 > 0$

So (3, –27) is a minimum stationary point. — Essential statement giving coordinates and nature of the point.

1 For each nature table:

i do the calculations necessary to support the results in the table

ii state the reason why the x-value gives a stationary point

iii state the coordinates and nature of the stationary point.

a $f(x) = 6x - x^2$

x	3^-	3	3^+
$f'(x)$	+	0	–
Shape	/	—	\

b $f(x) = 3 - x^3$

x	0^-	0	0^+
$f'(x)$	–	0	–
Shape	\	—	\

c $f(x) = x^2 - 14x + 49$

x	7^-	7	7^+
$f'(x)$	–	0	+
Shape	\	—	/

d $f(x) = 4x^3 - x^4$

x	0^-	0	0^+
$f'(x)$	+	0	+
Shape	/	—	/

e $f(x) = x(x-3)^2$

x	3^-	3	3^+
$f'(x)$	$-$	0	$+$
Shape	\	—	/

f $f(x) = x(x-3)^2$

x	1^-	1	1^+
$f'(x)$	$+$	0	$-$
Shape	/	—	\

g $f(x) = x^2(4-x^2)$

x	$\sqrt{2}^-$	$\sqrt{2}$	$\sqrt{2}^+$
$f'(x)$	$+$	0	$-$
Shape	/	—	\

h $f(x) = \sqrt{10 - x^2}$

x	0^-	0	0^+
$f'(x)$	$+$	0	$-$
Shape	/	—	\

2 Find algebraically the coordinates of the stationary points and determine their nature for these curves:

a $y = x^3 - 9x^2 + 2$

b $y = x^4 + 4x^3 - 1$

c $y = 2x^3 + 3x^2 - 12x + 2$

d $y = 3x^4 - 8x^3 + 6x^2$

e $y = 4x^5 + 5x^4 - 2$

f $y = 3x^4 - 4x^3 - 36x^2 + 5$

g $y = 3x^5 - 15x^4 + 20x^3$

h $y = 6x^4 + 8x^3 + 3x^2 + \frac{7}{8}$

3 Find algebraically the values of x for which the function:

a $f(x) = 6x^2 - x^3$ is increasing

b $f(x) = x^3 - 3x^2 - 9x$ is decreasing

c $g(x) = x^3 - 9x^2 + 15x$ is increasing

d $h(x) = 24x - 9x^2 - 2x^3$ is decreasing

e $f(x) = x^3 + 2x^2 + x$ is stationary or decreasing

f $g(x) = 4x + \frac{9}{2}x^2 - 3x^3$ is stationary or increasing

g $h(x) = -2x^3 + \frac{11}{2}x^2 - 4x$ is decreasing

h $f(x) = 5x^3 + 7x^2 - 8x$ is increasing or stationary

> **Hint** Make up a table of signs and use it to find the required range of values of x.

4 Determine the stationary points and their nature for the function $f(x) = x^3(3x^2 - 5)$.

5 Find all the stationary points on the graph $y = x^4 - 8x^2 + 16$ and determine the nature of each.

6 For each of these curves:

i find where the curve meets the x- and y-axes

ii find the stationary points and determine their nature

iii sketch the curve.

a $y = (x+2)^2(x-1)$

b $y = 3x - x^3$

11 Integrating functions

Exercise 11.1 Integration: the power rule

Example

Find $\int\left(2\sqrt{x}-\frac{4}{5\sqrt{x}}\right)dx$

$\int\left(2\sqrt{x}-\frac{4}{5\sqrt{x}}\right)dx$ — This is not in a form that can be integrated.

$=\int\left(2x^{\frac{1}{2}}-\frac{4}{5x^{\frac{1}{2}}}\right)dx$ — Write the roots as powers of x.

$=\int\left(2x^{\frac{1}{2}}-\frac{4x^{-\frac{1}{2}}}{5}\right)dx$ — Use the rule: $x^{-n}=\frac{1}{x^n}$ to bring $x^{\frac{1}{2}}$ to the top.

$=2\int x^{\frac{1}{2}}\,dx-\frac{4}{5}\int x^{-\frac{1}{2}}\,dx$ — Integrate each term separately.

$=\frac{2}{1}\times\frac{x^{\frac{3}{2}}}{\frac{3}{2}}-\frac{4}{5}\times\frac{x^{\frac{1}{2}}}{\frac{1}{2}}+c$ — Use the rule: $\int x^n\,dx=\frac{x^{n+1}}{n+1}+c$ to integrate each term.

$=\frac{2}{1}\times\frac{x^{\frac{3}{2}}\times 2}{\frac{3}{2}\times 2}-\frac{4}{5}\times\frac{x^{\frac{1}{2}}\times 2}{\frac{1}{2}\times 2}+c$ — Double top and bottom to get rid of $\frac{3}{2}$ and $\frac{1}{2}$.

$=\frac{4x^{\frac{3}{2}}}{3}-\frac{8x^{\frac{1}{2}}}{5}+c$ — Rewrite using root signs.

$=\frac{4\left(\sqrt{x}\right)^3}{3}-\frac{8\sqrt{x}}{5}+c$

$=\frac{4}{3}\left(\sqrt{x}\right)^3-\frac{8}{5}\sqrt{x}+c$ — Either of these is a suitable answer.

1 Find:

a $\int x^3\,dx$

b $\int 4x\,dx$

c $\int 5x^4\,dx$

d $\int 6x^2\,dx$

e $\int x^2\,dx$

f $\int -3x^2\,dx$

g $\int -10x^4\,dx$

h $\int 5x^2\,dx$

i $\int x\,dx$

j $\int(2x-4x^2)\,dx$

k $\int(6x^2-6x)\,dx$

l $\int(x+3x^2)\,dx$

m $\int(4x^3-5x^4)\,dx$

n $\int(9x^2+2x)\,dx$

o $\int(5x^9+3x^5-x)\,dx$

p $\int(3x-x^2-2x^3)\,dx$

Hint The rule $\int x^n\,dx=\frac{x^{n+1}}{n+1}+c$ works with x which is x^1 so use $n=1$.

2 Find:

a $\int 5\,dx$ **b** $\int -2\,dx$ **c** $\int \frac{1}{2}\,dx$

d $\int (4-x)\,dx$ **e** $\int (2x-1)\,dx$ **f** $\int (3x^2+2)\,dx$

g $\int (2x^2-x-1)\,dx$ **h** $\int (-4x-2)\,dx$ **i** $\int (-6x^4+2x)\,dx$

Hint: $\int a\,dx = ax + c$ where a is a constant.

3 Find:

a $\int \frac{1}{x^2}\,dx$ **b** $\int \frac{5}{x^3}\,dx$ **c** $\int -\frac{3}{x^2}\,dx$

d $\int \frac{5}{2x^2}\,dx$ **e** $\int \frac{1}{3x^4}\,dx$ **f** $\int -\frac{2}{5x^3}\,dx$

4 Find:

a $\int \left(x - \frac{2}{x^2}\right) dx$ **b** $\int \left(\frac{1}{x^3} - 2x^2\right) dx$ **c** $\int \left(5x - \frac{5}{x^6}\right) dx$

d $\int \left(\frac{4}{x^3} + \frac{3}{x^4}\right) dx$ **e** $\int \left(\frac{3}{2x^2} + \frac{5}{3x^6}\right) dx$ **f** $\int \left(5x^2 + 5 - \frac{3}{2x^2}\right) dx$

5 Find:

a $\int x^{\frac{1}{2}}\,dx$ **b** $\int 3x^{\frac{1}{2}}\,dx$ **c** $\int 2x^{-\frac{1}{2}}\,dx$

d $\int \frac{5x^{\frac{1}{2}}}{2}\,dx$ **e** $\int \frac{x^{-\frac{1}{2}}}{2}\,dx$ **f** $\int \frac{1}{3x^{\frac{1}{2}}}\,dx$

6 Integrate with respect to x:

a $x(x-1)$ **b** $x^2(3-x)$ **c** $(x-3)(2x+1)$

d $-3x(x-2)$ **e** $(x-3)(x+1)$ **f** $(2x+1)(x^2-1)$

g $\sqrt{x}\left(\frac{1}{\sqrt{x}} - \sqrt{x}\right)$ **h** $\sqrt{x}\left(\sqrt{x}-1\right)$ **i** $x\left(\sqrt{x}-1\right)$

7 Find:

a $\int 5\sqrt{x}\,dx$ **b** $\int \frac{4}{\sqrt{x}}\,dx$ **c** $\int \left(x - \frac{1}{\sqrt{x}}\right) dx$

d $\int \left(\frac{2}{3\sqrt{x}} + \sqrt{x}\right) dx$ **e** $\int \left(6\sqrt{x} - \frac{6}{\sqrt{x}}\right) dx$ **f** $\int x\left(1-\sqrt{x}\right) dx$

8 Integrate with respect to the given variable:

a $\sqrt{t}$ **b** $m^{-\frac{1}{3}}$ **c** $\frac{1}{\sqrt{n}}$ **d** $\sqrt[3]{x}$

e $-\frac{2}{\sqrt[3]{y}}$ **f** $h^{-\frac{2}{3}}$ **g** $k^{\frac{4}{3}}$ **h** $4w^{\frac{2}{3}}$

i $u^{-\frac{3}{4}}$ **j** $7v^{\frac{2}{5}}$ **k** $\frac{2}{3\sqrt[3]{g}}$ **l** $-\frac{5}{x^{\frac{1}{4}}}$

Example

Find $\int \frac{(2x^2-3)(2x^2+3)}{2x^2}\,dx$

$\int \frac{(2x^2-3)(2x^2+3)}{2x^2}\,dx$

Simplify the numerator.

$= \int \left(\frac{4x^4-9}{2x^2}\right)dx$

Now split the fraction into two separate fractions.

$= \int \left(\frac{4x^4}{2x^2} - \frac{9}{2x^2}\right)dx$

Do any cancelling and use the index rules: $\frac{x^m}{x^n} = x^{m-n}$ $\frac{1}{x^n} = x^{-n}$

$= \int \left(2x^2 - \frac{9x^{-2}}{2}\right)dx$

Integrate each term separately.

$= 2\int x^2\,dx - \frac{9}{2}\int x^{-2}dx$

Use: $\int x^n\,dx = \frac{x^{n+1}}{n+1} + c$

$= 2 \times \frac{x^3}{3} - \frac{9}{2} \times \frac{x^{-1}}{-1} + c$

Don't forget the constant c.

$= \frac{2}{3}x^3 + \frac{9}{2x} + c$

Tidy up the final answer.

9 Integrate with respect to x:

a $\frac{x^4-x^2}{x^2}$ **b** $\frac{x^4-3x}{x}$ **c** $\frac{x-1}{x^3}$

d $\frac{2x^3-1}{2x}$ **e** $\frac{x^3+x}{x}$ **f** $\frac{x^4+1}{x^2}$

g $\frac{(x^2-1)^2}{x^2}$ **h** $\frac{x^2+3}{\sqrt{x}}$ **i** $\frac{x(3x-2)}{\sqrt{x}}$

j $\frac{5(x-2)}{\sqrt[3]{x}}$ **k** $\frac{8(x-1)(x+1)}{\sqrt[3]{x}}$ **l** $\frac{1-x}{x\sqrt{x}}$

Exercise 11.2 Integrating powers of linear functions

Example

$\int \frac{3}{4\sqrt{5x-1}}\,dx$

$\int \frac{3}{4\sqrt{5x-1}}\,dx$

Change the root to a power.

$= \int \frac{3}{4(5x-1)^{\frac{1}{2}}}\,dx$

Use $\frac{1}{x^n} = x^{-n}$ with $(5x-1)^{\frac{1}{2}}$

$= \frac{3}{4}\int (5x-1)^{-\frac{1}{2}}\,dx$

Use $\int (ax+b)^n\,dx = \frac{(ax+b)^{n+1}}{a(n+1)} + c$ with $a = 5$, $b = -1$, $n = -\frac{1}{2}$

$= \frac{3}{4} \times \frac{(5x-1)^{\frac{1}{2}}}{5 \times \frac{1}{2}} + c$

$= \frac{3}{10}\sqrt{5x-1} + c$

Tidy up the answer.

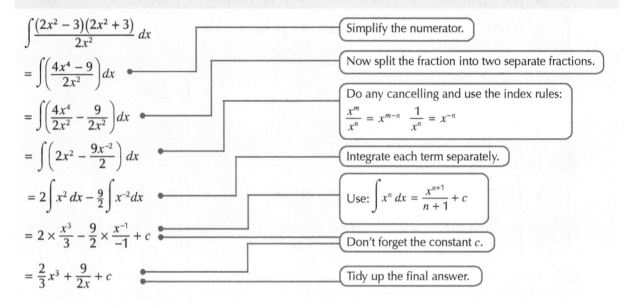

1 Find:

a $\int(2x-1)^5\,dx$ **b** $\int(3x+1)^4\,dx$ **c** $\int(1-2x)^3\,dx$

d $\int(1-x)^6\,dx$ **e** $\int\left(\frac{1}{2}x+1\right)^3\,dx$ **f** $\int\left(\frac{1}{3}x-1\right)^4\,dx$

g $\int\left(1-\frac{1}{2}x\right)^3\,dx$ **h** $\int\left(2-\frac{1}{3}x\right)^5\,dx$ **i** $\int\left(\frac{2}{3}x+2\right)^3\,dx$

> **Hint** $\int(ax+b)^n\,dx = \frac{(ax+b)^{n+1}}{a(n+1)}+c$

2 Here are some possible results from integrating $(2x+1)^k$.
Match each result with the correct value of k.

Ⓐ $\frac{1}{4}(2x+1)^2+c$

Ⓑ $-(2x+1)^{-\frac{1}{2}}+c$

Ⓒ $(2x+1)^{\frac{1}{2}}+c$

Ⓓ $\frac{1}{3}(2x+1)^{\frac{3}{2}}+c$

Ⓔ $-\frac{1}{3}(2x+1)^{-\frac{3}{2}}+c$

Ⓕ $\frac{1}{5}(2x+1)^{\frac{5}{2}}+c$

① $k=-\frac{3}{2}$

② $k=-\frac{5}{2}$

③ $k=-\frac{1}{2}$

④ $k=\frac{1}{2}$

⑤ $k=\frac{3}{2}$

⑥ $k=1$

3 Find:

a $\int(3x-1)^2\,dx$ **b** $\int(5x+1)^3\,dx$ **c** $\int(3-2x)^2\,dx$

d $\int(4+6x)^4\,dx$ **e** $\int2(3x+4)^3\,dx$ **f** $\int3(5-2x)^2\,dx$

g $\int(2-x)^4\,dx$ **h** $\int6(7-x)^3\,dx$ **i** $\int-(1-3x)^2\,dx$

4 Integrate with respect to x:

a $2(2x-1)^5$ **b** $3(1-4x)^3$ **c** $-2(1-x)^4$

d $4(3-2x)^3$ **e** $-(3x+1)^5$ **f** $\frac{1}{2}(2x+3)^3$

g $\frac{3}{2}(3x+1)^7$ **h** $-\frac{1}{2}(2x-3)^3$ **i** $-\frac{3}{4}\left(\frac{1}{2}x-1\right)^2$

5 Here are some possible results from integrating $k(4x-1)^{-2}$.
Match each result with the correct value of k.

Ⓐ $-\frac{1}{8}(4x-1)^{-1}+c$

Ⓑ $\frac{1}{4}(4x-1)^{-1}+c$

Ⓒ $\frac{1}{8}(4x-1)^{-1}+c$

Ⓓ $-(4x-1)^{-1}+c$

Ⓔ $\frac{1}{2}(4x-1)^{-1}+c$

Ⓕ $-\frac{1}{2}(4x-1)^{-1}+c$

① $k=2$

② $k=-2$

③ $k=\frac{1}{2}$

④ $k=4$

⑤ $k=-\frac{1}{2}$

⑥ $k=-1$

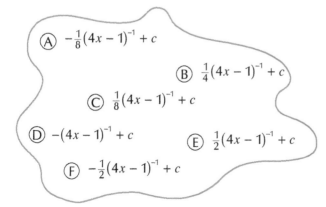

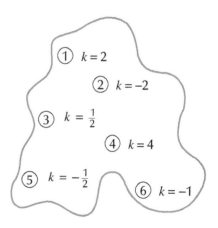

6 Find:

a $\int (2x+3)^{-2}\,dx$ b $\int (5-x)^{-3}\,dx$ c $\int (3-2x)^{-5}\,dx$

d $\int (4x-2)^{-2}\,dx$ e $\int \left(\frac{1}{2}x+1\right)^{-3}\,dx$ f $\int \left(\frac{1}{3}x-2\right)^{-4}\,dx$

7 Integrate with respect to x:

a $(1-x)^{\frac{1}{2}}$ b $(2x-3)^{-\frac{1}{2}}$ c $(4x-1)^{\frac{3}{2}}$ d $(3x-1)^{\frac{1}{3}}$ e $(2-5x)^{-\frac{1}{5}}$

f $(3-2x)^{-\frac{1}{2}}$ g $(6x+5)^{\frac{2}{3}}$ h $(3x+2)^{-\frac{1}{3}}$ i $(5-x)^{-\frac{1}{4}}$

Exercise 11.3 Integrating some trigonometric functions

1 Integrate with respect to x:

a $\sin x$ b $\cos x$ c $-\sin x$ d $-\cos x$

e $2\cos x$ f $\frac{1}{2}\sin x$ g $-3\cos x$ h $-\frac{3}{4}\sin x$

i $1-\cos x$ j $\cos x+\sin x$ k $2\sin x-\cos x$

l $x+\cos x$ m $2\sin x-2x$ n $3+2\sin x+4\cos x$

2 Find:

a $\int (5x^3+x+\sin x)\,dx$ b $\int (4x-\cos x)\,dx$

c $\int (3x^2-2x-\sin x)\,dx$ d $\int (6x^3-6x^2+\cos x)\,dx$

e $\int (10x^4+2\cos x)\,dx$ f $\int (5x^2-x-3\sin x)\,dx$

g $\int (8x^3-9x^2-2\cos x)\,dx$ h $\int (10x+12x^3+4\sin x)\,dx$

3 Find:

a $\int \sin 2x\,dx$ b $\int \cos \frac{1}{2}x\,dx$ c $\int \sin\left(x-\frac{\pi}{6}\right)dx$

d $\int \cos\left(\frac{\pi}{2}-x\right)dx$ e $\int \cos(\pi-x)\,dx$ f $\int \cos\left(\frac{\pi}{2}-2x\right)dx$

g $\int \sin\left(2x+\frac{\pi}{3}\right)dx$ h $\int 2\cos\left(\frac{1}{2}x-\frac{\pi}{3}\right)dx$ i $\int \frac{1}{2}\cos\left(4x-\frac{\pi}{8}\right)dx$

Hint

$f(x)$	$\int f(x)dx$
$\sin ax$	$-\frac{1}{a}\cos ax+c$
$\cos ax$	$\frac{1}{a}\sin ax+c$

4 The double angle formulae can be arranged to give:

$$\sin^2 x=\frac{1}{2}-\frac{1}{2}\cos 2x \text{ and } \cos^2 x=\frac{1}{2}+\frac{1}{2}\cos 2x$$

Use these to find:

a $\int \sin^2 x\,dx$ b $\int \cos^2 x\,dx$

c Use your answers to parts **a** and **b** to show that $\int (\sin^2 x+\cos^2 x)\,dx=x+c$

d Explain an easier way to find $\int (\sin^2 x+\cos^2 x)\,dx$

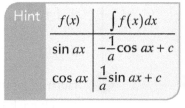

Exercise 11.4 Solving simple differential equations

1 In each case find the value of c.

a $P(-1, 2)$ lies on $y = x^2 - 2x + c$

b $P(2, -3)$ lies on $y = x^3 - x + c$

c $P(\frac{1}{2}, 0)$ lies on $y = \sqrt{2x+3} + c$

d $P(0, -\frac{1}{2})$ lies on $y = \dfrac{2}{4-x} + c$

e $P(-1, -1)$ lies on $y = (x+2)^7 + c$

f $P(\pi, 2)$ lies on $y = \cos x + c$

> **Hint** If (a, b) lies on $y = f(x)$, then $b = f(a)$, i.e. the coordinates satisfy the equation.

2 Use the information given in each diagram to find the equation of the curve.

a

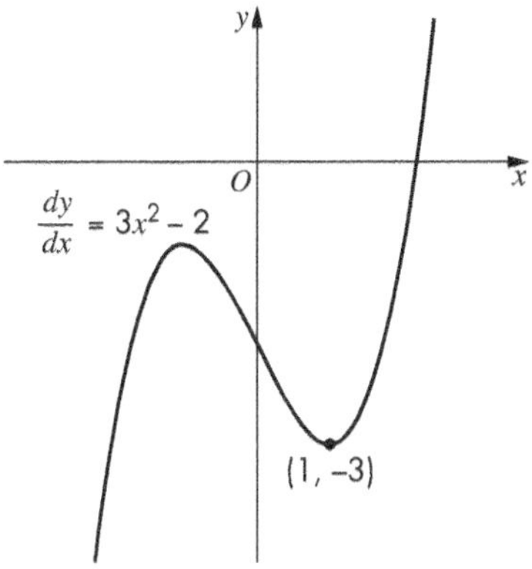

b

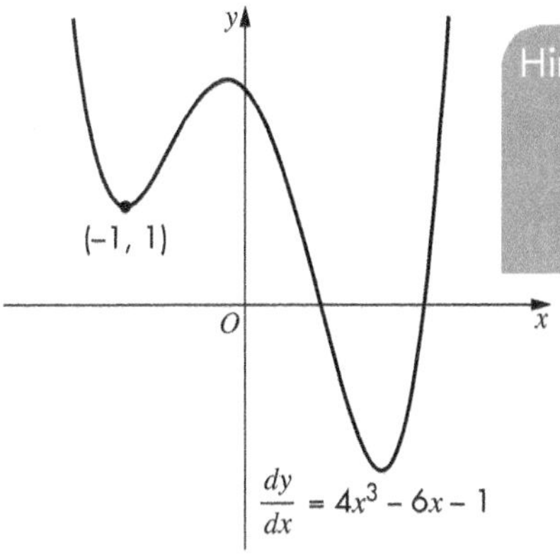

> **Hint** To find y from $\dfrac{dy}{dx}$ integrate.

3 In each case find an expression for $f(x)$.

a $f'(x) = 2\sqrt{x}$ and $f(1) = 1$

b $f'(x) = \dfrac{3}{x^2}$ and $f(3) = -5$

c $f'(x) = \cos x - \sin x$ and $f\left(\dfrac{\pi}{2}\right) = -2$

4 The diagram shows five curves, all of which satisfy the differential equation $\dfrac{dy}{dx} = x$.

A point is shown lying on each curve.

For each point find the equation of the curve that it lies on.

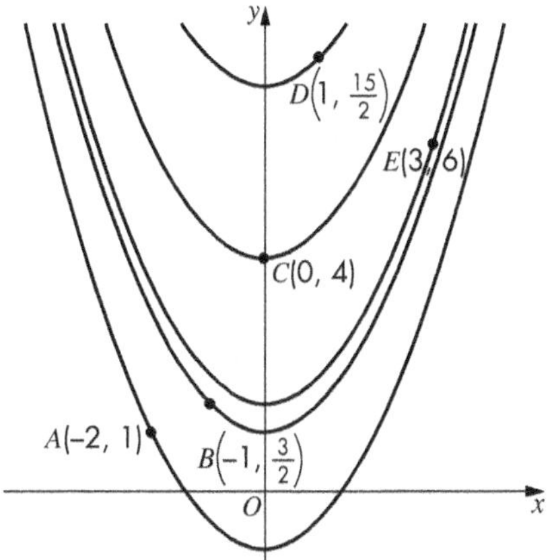

5 Use the information given in each diagram to find the equation of the curve.

a

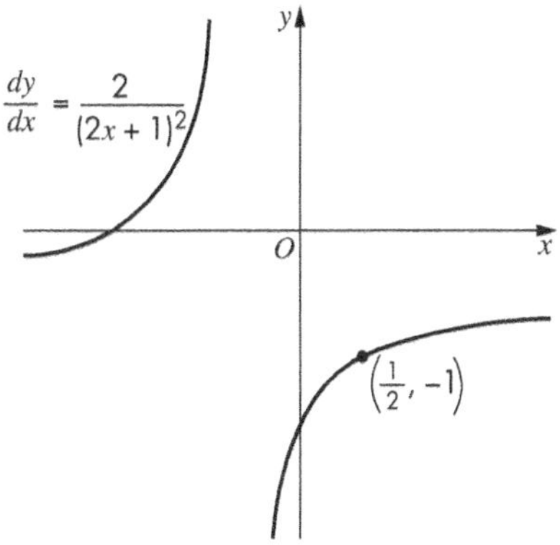

b

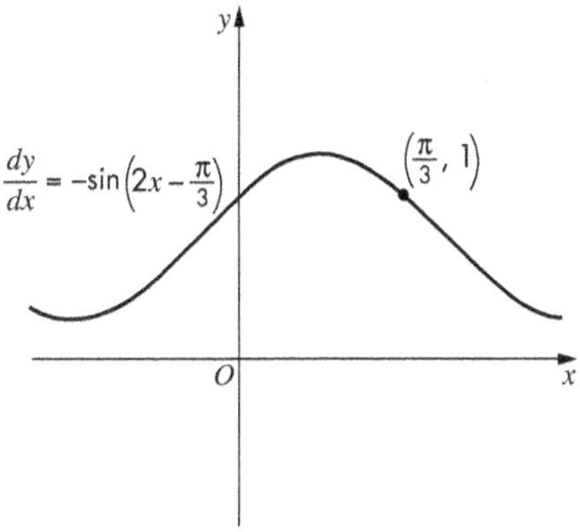

12 Using integration to calculate definite integrals

Exercise 12.1 Evaluating definite integrals

Example

Evaluate $\int_{-1}^{1}(2x-3)\,dx$

$\int_{-1}^{1}(2x-3)\,dx$ — Integrate $2x-3$.

$= [x^2 - 3x]_{-1}^{1}$ — Check the result by differentiating $x^2 - 3x$ to get $2x - 3$. Place the result in: $[\text{square brackets}]_{\text{lower limit}}^{\text{upper limit}}$

$= (1^2 - 3\times 1) - ((-1)^2 - 3\times(-1))$ — Use $\begin{pmatrix}\text{expression evaluated}\\ \text{at the upper limit}\end{pmatrix} - \begin{pmatrix}\text{expression evaluated}\\ \text{at the lower limit}\end{pmatrix}$

$= (1-3) - (1+3)$ — Take great care with negative signs.

$= -2 - 4 = -6$ — Simplify and calculate the final result.

1 Evaluate:

a $\int_{0}^{1} 1\,dx$ **b** $\int_{-1}^{1} 1\,dx$ **c** $\int_{2}^{3} 1\,dx$

d $\int_{0}^{2} -1\,dx$ **e** $\int_{0}^{1} x\,dx$ **f** $\int_{-1}^{1} x\,dx$

g $\int_{-1}^{0} x\,dx$ **h** $\int_{1}^{3} 2x\,dx$ **i** $\int_{-1}^{1} 2x\,dx$

j $\int_{0}^{1}(2x-1)\,dx$ **k** $\int_{-1}^{1}(4x-1)\,dx$ **l** $\int_{-1}^{0}(2x-1)\,dx$

m $\int_{1}^{2}(6x-1)\,dx$ **n** $\int_{-2}^{-1}(2x-1)\,dx$ **o** $\int_{2}^{3}(2x-1)\,dx$

p $\int_{-1}^{3} 5\,dx$ **q** $\int_{3}^{5} 6x\,dx$ **r** $\int_{-2}^{1} -3\,dx$

s $\int_{1}^{2}(1-2x)\,dx$ **t** $\int_{-2}^{-1}(1-2x)\,dx$ **u** $\int_{-3}^{0}(2-4x)\,dx$

2 Find expressions in terms of a or b for:

a $\int_{a}^{b} 1\,dx$ **b** $\int_{a}^{b} 3\,dx$ **c** $\int_{a}^{2a} 5\,dx$

d $\int_{0}^{a} 2\,dx$ **e** $\int_{-a}^{0} 2\,dx$ **f** $\int_{-a}^{a} 2\,dx$

g $\int_{1}^{a} 7\,dx$ **h** $\int_{a}^{1} 7\,dx$ **i** $\int_{a}^{b} -4\,dx$

j $\int_{a}^{b}(2x-1)\,dx$ **k** $\int_{a}^{b}(1-2x)\,dx$ **l** $\int_{-a}^{a}(2x-1)\,dx$

3 Evaluate:

a $\int_{0}^{2} x\,dx$ **b** $\int_{1}^{2} 3\,dx$ **c** $\int_{-1}^{1}(2x-1)\,dx$

d $\int_{-2}^{1}(5-3x)\,dx$ **e** $\int_{1}^{3} 3x^2\,dx$ **f** $\int_{0}^{1}(x^2-x+2)\,dx$

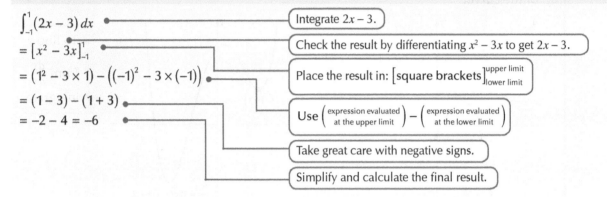

4 Find the exact value of:

a $\int_1^2 (3x^2 - 2x - 1)\,dx$ b $\int_0^3 (4x^3 - 3)\,dx$

c $\int_{-1}^1 (3x^2 + 2)\,dx$ d $\int_{-2}^0 (5x^4 + 2x)\,dx$

e $\int_{-2}^{-1} (2x - 1)\,dx$ f $\int_{-2}^{-1} (3x^2 + 2x)\,dx$

g $\int_{-1}^1 (x - 1)(x + 1)\,dx$ h $\int_1^2 x(x - 1)\,dx$

Hint

$f(x)$	$\int f(x)dx$
$\sin ax$	$-\frac{1}{a}\cos ax + c$
$\cos ax$	$\frac{1}{a}\sin ax + c$

5 Show that the following results are true.

a $\int_0^a 1\,dx = \int_a^0 -1\,dx$

b $\int_a^b 1\,dx = \int_b^a -1\,dx$

c $\int_a^b 1\,dx + \int_b^c 1\,dx = \int_a^c 1\,dx$

d $\int_0^b 1\,dx - \int_0^a 1\,dx = \int_a^b 1\,dx$

e $\int_{-a}^a 1\,dx = \int_0^a 2\,dx$

f $\int_{-a}^a (1 - 2x)\,dx = \int_0^a 2\,dx$

Hint Find an expression for the left-hand side.

Find an expression for the right-hand side.

Are they equal?

6 Match the integrals and values.

Ⓐ $\int_0^1 \sqrt{x}\,dx$

Ⓑ $\int_0^1 3\sqrt{x}\,dx$

Ⓒ $\int_0^4 \frac{1}{\sqrt{x}}\,dx$

Ⓓ $\int_4^{25} \frac{1}{2\sqrt{x}}\,dx$

Ⓔ $\int_1^4 \frac{3}{\sqrt{x}}\,dx$

① 4

② 6

③ $\frac{2}{3}$

④ 3

⑤ 2

7 Find the exact value of:

a $\int_9^{25} \frac{1}{\sqrt{x}}\,dx$ b $\int_4^9 3\sqrt{x}\,dx$ c $\int_1^9 \frac{3}{2\sqrt{x}}\,dx$

d $\int_4^9 \left(\sqrt{x} - \frac{1}{\sqrt{x}}\right)dx$ e $\int_0^4 \sqrt{x}\left(\sqrt{x} - 1\right)dx$ f $\int_1^4 \left(\sqrt{x} + \frac{1}{\sqrt{x}}\right)dx$

g $\int_1^9 \left(\frac{1}{x^2} + \sqrt{x}\right)dx$ h $\int_1^4 \left(\frac{4}{x^3} + \frac{1}{2\sqrt{x}}\right)dx$

8 Evaluate:

a $\int_1^3 x^2\,dx$ b $\int_1^2 (2x - 1)^2\,dx$ c $\int_{-1}^1 (3x + 1)^3\,dx$

d $\int_2^4 \left(\frac{1}{2}x - 1\right)^3 dx$ e $\int_{-1}^0 (5x - 1)^2\,dx$ f $\int_{-\frac{1}{2}}^0 (1 - 2x)^4\,dx$

g $\int_{-3}^0 \left(\frac{1}{3}x - 2\right)^2 dx$ h $\int_{-4}^0 \left(\frac{1}{4}x + 1\right)^7 dx$ i $\int_{-2}^2 \left(1 - \frac{1}{2}x\right)^3 dx$

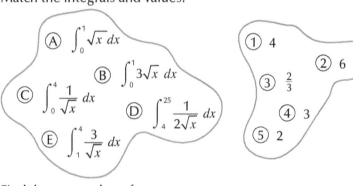

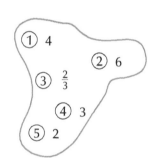

Exercise 12.2 Evaluating definite trigonometric integrals

1 Evaluate:

a $\int_0^{\pi} \sin x \, dx$ **b** $\int_0^{\frac{\pi}{2}} \cos x \, dx$

c $\int_{\frac{\pi}{2}}^{\pi} \cos x \, dx$ **d** $\int_{\pi}^{2\pi} \sin x \, dx$

e $\int_0^{\frac{3\pi}{2}} \sin x \, dx$ **f** $\int_0^{\frac{\pi}{4}} \cos 2x \, dx$

g $\int_0^{\frac{\pi}{2}} \sin 2x \, dx$ **h** $\int_{\frac{\pi}{4}}^{\frac{\pi}{2}} \sin 2x \, dx$

i $\int_{\frac{\pi}{2}}^{\frac{3\pi}{4}} \cos 2x \, dx$ **j** $\int_{\pi}^{2\pi} \cos \tfrac{1}{2}x \, dx$

Hint Remember:

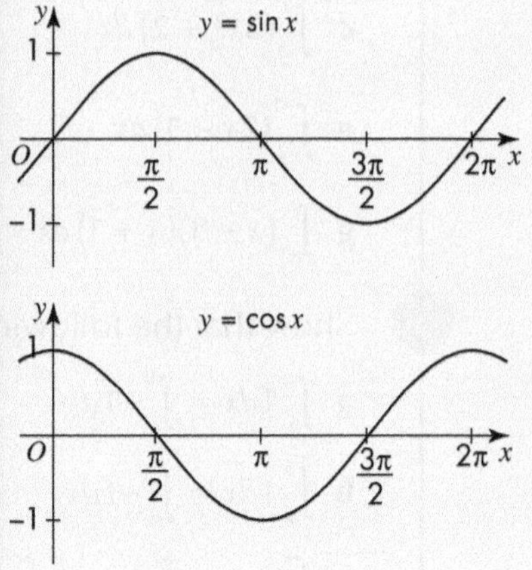

2 There is a mistake in each of these calculations. Find it and fix it.

a $\displaystyle\int_0^{\frac{\pi}{3}} \cos x \, dx = \Big[-\sin x\Big]_0^{\frac{\pi}{3}} = -\sin\frac{\pi}{3} - (-\sin 0) = -\frac{\sqrt{3}}{2} + 0 = -\frac{\sqrt{3}}{2}$

b $\displaystyle\int_0^{\frac{\pi}{3}} \sin x \, dx = \Big[-\cos x\Big]_0^{\frac{\pi}{3}} = -\cos\frac{\pi}{3} - (-\cos 0) = -\frac{1}{2} + 0 = -\frac{1}{2}$

c $\displaystyle\int_0^{\frac{\pi}{4}} 2\cos x \, dx = \Big[2\sin x\Big]_0^{\frac{\pi}{4}} = 2\sin\frac{\pi}{4} - 2\sin 0 = 2\sqrt{2} - 0 = 2\sqrt{2}$

d $\displaystyle\int_0^{\frac{\pi}{4}} \sqrt{2}\sin x \, dx = \Big[-\sqrt{2}\cos x\Big]_0^{\frac{\pi}{4}} = -\sqrt{2}\cos\frac{\pi}{4} - (-\sqrt{2}\cos 0) = -1 - \sqrt{2}$

3 Find the exact value of the following integrals.
Write your answer as a single fraction where necessary.

a $\int_0^{\frac{\pi}{4}} \cos x \, dx$ **b** $\int_{\frac{\pi}{3}}^{\frac{\pi}{2}} \sin x \, dx$

c $\int_0^{\frac{\pi}{4}} (\sin x + \cos x) dx$ **d** $\int_{\frac{\pi}{3}}^{\frac{\pi}{2}} \cos x \, dx$

e $\int_0^{\frac{\pi}{6}} \sin x \, dx$ **f** $\int_{\frac{\pi}{4}}^{\frac{\pi}{2}} (\cos x - \sin x) \, dx$

Hint

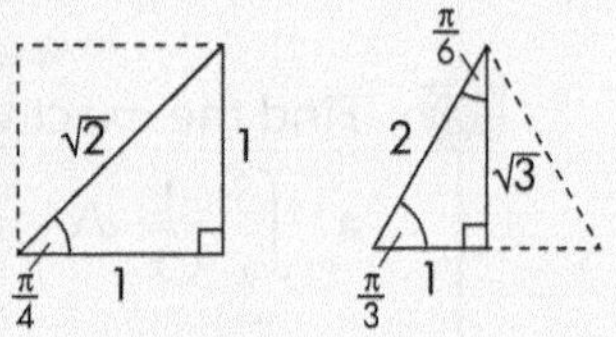

4 Evaluate:

a $\displaystyle\int_0^{\frac{\pi}{3}} \cos 3x \, dx$ **b** $\displaystyle\int_0^{\frac{\pi}{4}} \sin 4x \, dx$

c $\displaystyle\int_{-\frac{\pi}{4}}^{\frac{\pi}{4}} \cos\left(5x - \frac{\pi}{4}\right) dx$ **d** $\displaystyle\int_{-\frac{\pi}{6}}^{\frac{\pi}{6}} \sin\left(2x + \frac{\pi}{3}\right) dx$

5 **a** The double angle formula can be rearranged to: $\cos^2\theta = \frac{1}{2}\cos 2\theta + \frac{1}{2}$

Use this result to show that $\int_0^{\frac{\pi}{4}} \cos^2\theta \, d\theta$ has exact value $\dfrac{\pi + 2}{8}$

b Rearrange the formula $\cos 2\theta = 1 - 2\sin^2\theta$ to find a similar formula for $\sin^2\theta$ and then use it to find the exact value of $\int_0^{\frac{\pi}{4}} \sin^2\theta \, d\theta$.

13 Applying algebraic skills to rectilinear shapes

Exercise 13.1 Review exercise

1 Find the gradient of the line joining the given pairs of points.

a $A(3, 5), B(5, 7)$ b $P(1, -1), Q(3, 5)$ c $R(1, 1), S(3, -5)$

d $M(-3, -5), N(5, -1)$ e $O(0, 0), T(-1, 5)$ f $C(-7, -1), D(3, -6)$

g $V(7, -2), W(12, -2)$ h $G\left(1, -\frac{3}{2}\right), H(3, -1)$ i $E\left(\frac{1}{2}, \frac{1}{2}\right), F\left(-\frac{3}{2}, -1\right)$

Hint $m = \frac{y_2 - y_1}{x_2 - x_1}$

2 Decide in each case whether the given point lies on the line.

a $(2, 3); 3x + 2y = 12$ b $(-2, 1); 3x - 2y = -4$ c $(-2, -3); x - 3y = 7$

d $\left(\frac{1}{2}, 1\right); -2x + y = -1$ e $\left(-1, \frac{1}{2}\right); 3x = 2y - 4$ f $\left(-\frac{1}{2}, \frac{3}{2}\right); x + 2 = y$

Hint Do the coordinate values satisfy the equation?

3 Find the equation of the line passing through the given point with the given gradient.

a $(2, 3); 2$ b $(1, -3); \frac{1}{2}$ c $(-2, 5); -\frac{1}{2}$

d $(5, -1); -3$ e $(0, 4); -1$ f $\left(\frac{3}{2}, 0\right); \frac{1}{2}$

Hint (a, b) is a point on the line. m is the gradient of the line. The equation of the line is: $y - b = m(x - a)$

4 For each line find the points of intersection with the x- and y-axes.

a $y = 3x + 6$ b $2y - 3x = 6$ c $3y + x = 6$

d $2y - 5x - 10 = 0$ e $5y - 2x = 5$ f $3y + 2x + 5 = 0$

Hint For the x-axis intercept set $y = 0$ in the equation. For the y-axis intercept set $x = 0$ in the equation.

5 For each pair of lines find their point of intersection.

a $2x - y = -1$, $x + 2y = 12$ b $x + 3y - 8 = 0$, $3x + 2y - 3 = 0$ c $4x + 3y - 10 = 0$, $2x - y - 10 = 0$

d $6x - y - 3 = 0$, $2x + 3y - 1 = 0$ e $5x = 2y - 2$, $2x = 3y + 8$ f $2x + 4y + 5 = 0$, $y = -3x$

Hint Treat these as simultaneous equations.

6 Determine whether or not these lines are concurrent:

a $2x + y - 7 = 0$, $x + 2y - 11 = 0$ and $3x - y + 2 = 0$

b $3x - y + 2 = 0$, $2x + 3y + 5 = 0$ and $x - y + 1 = 0$

c $x + y - 1 = 0$, $5x + 2y + 4 = 0$ and $3x - y + 8 = 0$

d $2x - y = 0$, $6x + 5y = 8$ and $4x - 3y + 1 = 0$

Hint Lines are concurrent if they all pass through the same point.

7 Find the distance between these points:

a $(2, 1), (5, 5)$ b $(-5, 1), (1, 9)$ c $(5, -1), (1, -4)$

d $(10, 0), (-2, -5)$ e $\left(\frac{1}{2}, -1\right), \left(-\frac{1}{2}, 0\right)$ f $\left(\frac{3}{2}, -2\right), \left(-\frac{1}{2}, -6\right)$

g $(0, 0), (-2, 8)$ h $\left(-\frac{3}{2}, \frac{3}{2}\right), (-2, 1)$ i $\left(\sqrt{2}, \sqrt{2}\right), \left(0, 2\sqrt{2}\right)$

Hint $A(x_1, y_1)$ $B(x_2, y_2)$

$AB = \sqrt{(x_2 - x_1)^2 + (y_2 - y_1)^2}$

x-coordinate difference; y-coordinate difference

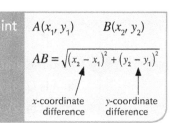

8 Find the coordinates of M, the midpoint of the line joining the given pair of points.

a (2, 3), (10, 1)
b (−2, 3), (4, 5)
c (−3, 1), (−1, −3)
d (−10, −1), (−2, −3)
e (1, 5), (4, −2)
f $\left(\frac{1}{2}, -\frac{3}{2}\right), \left(\frac{3}{2}, -\frac{1}{2}\right)$
g (0, 0), (−2, 7)
h $\left(-\frac{3}{2}, 0\right), \left(\frac{1}{2}, -3\right)$
i $\left(\sqrt{2}, \sqrt{2}\right), \left(-\sqrt{2}, \sqrt{2}\right)$

> **Hint** $A(x_1, y_1)$ $B(x_2, y_2)$
> $M\left(\frac{x_1 + x_2}{2}, \frac{y_1 + y_2}{2}\right)$
> 'average' of the x-coordinates; 'average' of the y-coordinates

Exercise 13.2 Parallel and perpendicular lines

1 In each case decide whether P lies on the line passing through A and B.

a $A(-2, 1)$, $B(1, 4)$; $P(4, 8)$
b $A(-2, 4)$, $B(-1, 2)$; $P(1, -2)$
c $A(5, 1)$, $B(-1, -2)$; $P(-3, -3)$
d $A(2, -4)$, $B(0, 2)$; $P(-2, 6)$
e $A\left(\frac{5}{2}, 0\right)$, $B\left(\frac{1}{2}, 1\right)$; $P\left(-\frac{3}{2}, 2\right)$
f $A\left(\sqrt{2}, 0\right)$, $B\left(0, -\sqrt{2}\right)$; $P\left(2\sqrt{2}, \sqrt{2}\right)$

> **Hint** Find m_{AB} and m_{BP}.

2 A radar operator notices an aircraft at $T(2, -1)$. After a few minutes she observes it at $U(-1, 2)$. There are airports at $A_1(-8, 8)$ and $A_2(-8, 9)$. For which airport is it heading if it does not change course?

3 Show that each triangle ABC is right-angled at B.

a

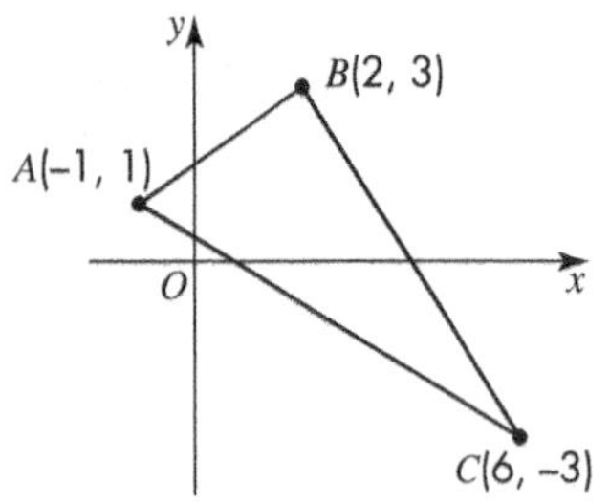

b

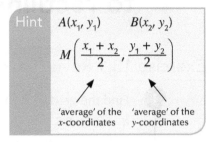

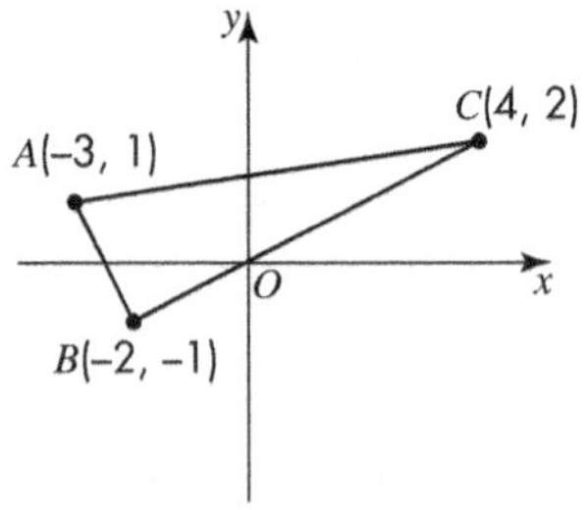

> **Hint** If $m_1 \times m_2 = -1$ then the two lines are perpendicular.

4 Show that each set of three points forms a right-angled triangle and name the right-angle in each case.

a $P(-8, 3)$, $Q(-3, 4)$, $R(-6, -7)$
b $C(-8, -7)$, $D(8, 4)$, $E(-2, 8)$
c $U(-7, 0)$, $V(-5, -7)$, $W(7, 4)$
d $A(-5, -1)$, $B(-9, -3)$, $C(-2, -7)$

5 Find the gradient of lines perpendicular to AB where:

a $m_{AB} = \frac{3}{4}$
b $m_{AB} = -\frac{2}{3}$
c $m_{AB} = 5$
d $m_{AB} = -2$
e $m_{AB} = 0$
f $m_{AB} = \frac{1}{3}$
g $m_{AB} = -\frac{1}{2}$
h $m_{AB} = -\frac{5}{2}$
i $m_{AB} = 1{\cdot}5$

6 Find the gradient of lines perpendicular to AB where:

a A is (2, 3), B is (4, 6)
b A is (−1, 3), B is (1, 8)
c A is (4, 2), B is (6, 1)
d A is (−3, 4), B is (−1, 1)

Exercise 13.3 Gradients and angles

Example

Calculate, to 1 decimal place, the angle $\theta°$ that the line passing through $A(2, 3)$ and $B(-1, 5)$ makes with the positive direction of the x-axis.

$A(2, 3)$, $B(-1, 5)$

$m_{AB} = \frac{5-3}{-1-2} = \frac{2}{-3} = -\frac{2}{3}$ — Use the gradient formula.

so $\tan\theta° = -\frac{2}{3}$ — Since the gradient is –ve, $\theta°$ is obtuse (2nd quadrant).

giving $\theta = 180 - 33{\cdot}69\ldots = 146{\cdot}30\ldots$ — Use $\tan^{-1}\frac{2}{3}$ to get the acute (1st quad) angle $33{\cdot}69\ldots°$

so $\theta \approx 146{\cdot}3$ (to 1 d.p.) — Now round to 1 decimal place.

1 Calculate the angle that the lines OA, OB, ... OG make with the positive direction of the x-axis. Give your answers in degrees correct to 1 decimal place.

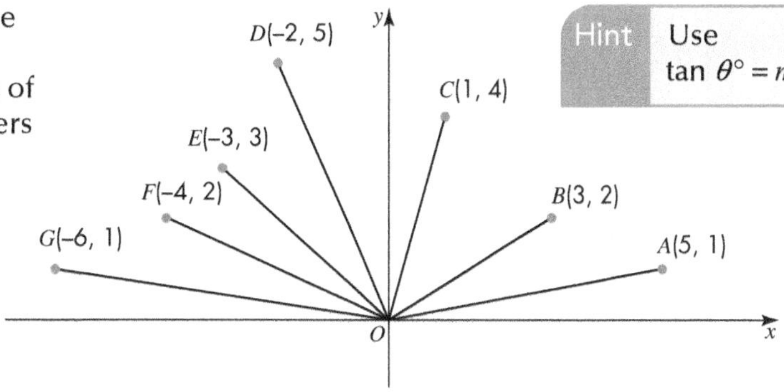

Hint Use $\tan\theta° = m$

2 Find, in degrees correct to 1 decimal place, the angle $\theta°$ which the line passing through A and B makes with the positive direction of the x-axis.

a $A(1, 5)$, $B(-1, 3)$ **b** $A(2, 8)$, $B(0, 2)$ **c** $A(2, 3)$, $B(-2, -3)$

d $A(11, -1)$, $B(-1, -5)$ **e** $A(7, 1)$, $B(4, 4)$ **f** $A(3, -1)$, $B(-1, 1)$

g $A(1, -1)$, $B(7, -3)$ **h** $A(-2, 3)$, $B\left(-1, \frac{3}{2}\right)$ **i** $A(-2, 3)$, $B\left(\frac{1}{2}, -4\right)$

3 Calculate, in degrees correct to 1 decimal place, the size of the three angles of each of these triangles ABC.

a

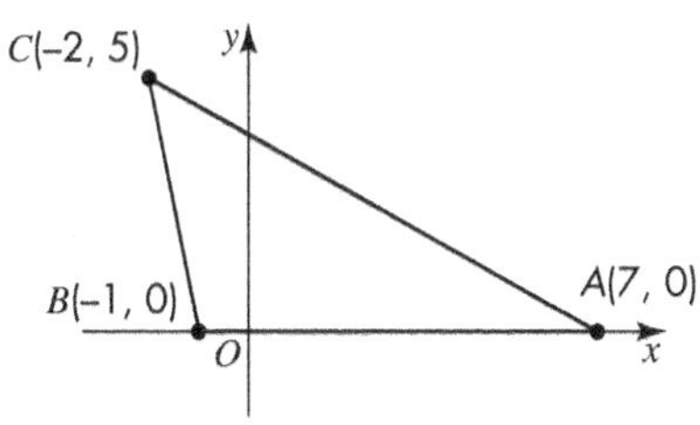

b

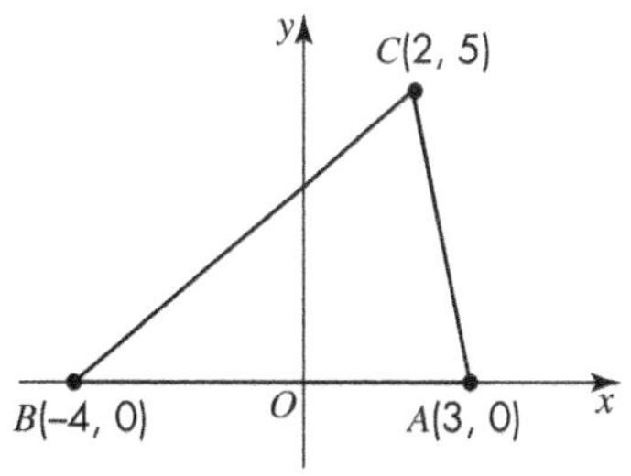

c

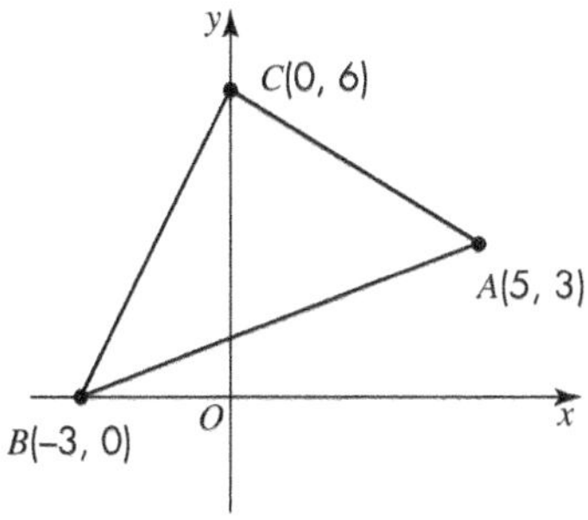

d

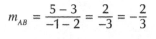

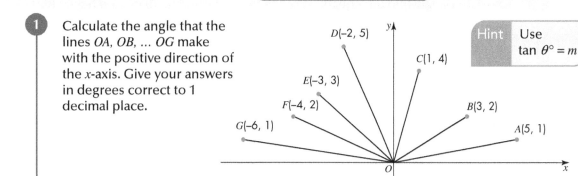

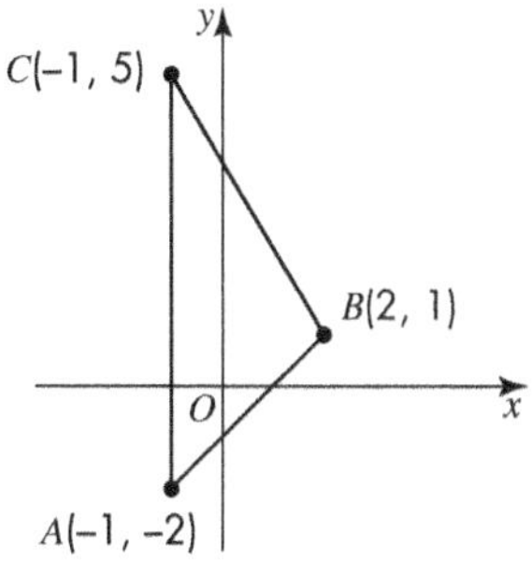

Exercise 13.4 Lines and triangles

1 Find the equation of the perpendicular bisector of the line joining:

a $A(-1, 1)$, and $B(3, 7)$ b $P(0, 4)$, and $Q(2, 2)$

c $R(-1, -1)$, and $S(3, 1)$ d $T(-8, 6)$, and $U(-2, 4)$

e $C(-1, -2)$, and $D(3, 0)$ f $E(0, 2)$, and $F(-2, 8)$

g $J(-2, -1)$, and $K(-4, 0)$ h $L(-3, -1)$, and $M(-7, 0)$

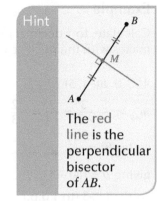

Hint

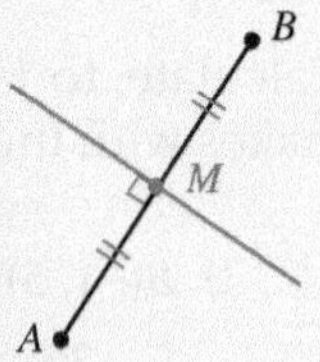

The red line is the perpendicular bisector of AB.

2 For each triangle find the equations of the two indicated medians and hence find G, their point of intersection.

a

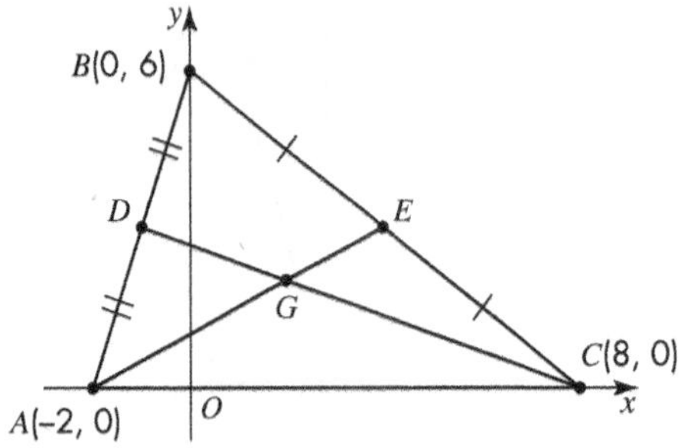

b

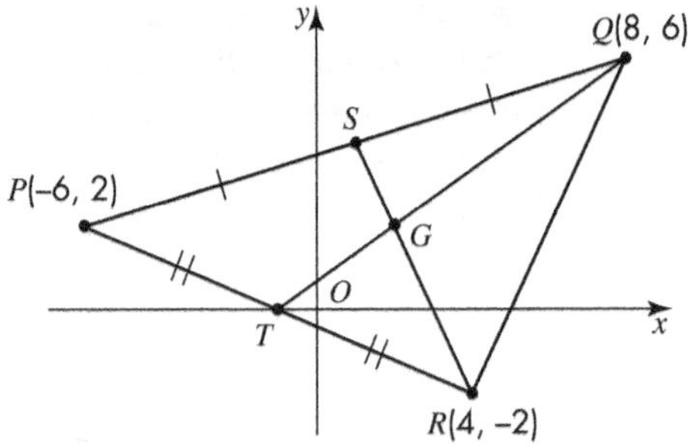

3 A triangle has vertices $A(-8, -1)$, $B(2, 1)$ and $C(0, -3)$. The medians BD and CE intersect at M.

a Find the equations of BD and CE.

b Hence find the coordinates of M.

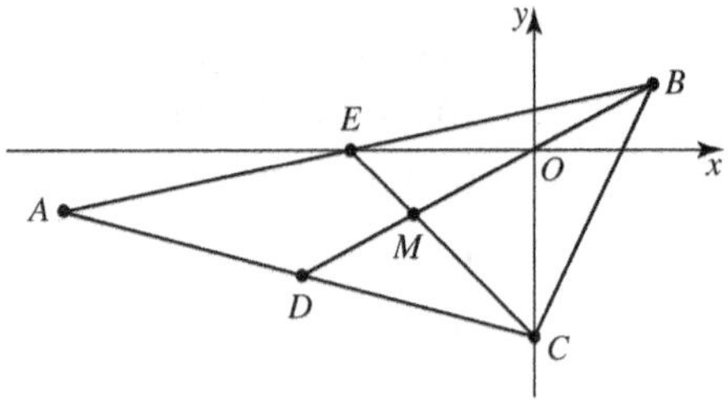

4 A triangle has vertices $P(-2, 3)$, $Q(6, -1)$ and $R(-4, -5)$. Find the coordinates of G, the point of intersection of the medians PS and RT.

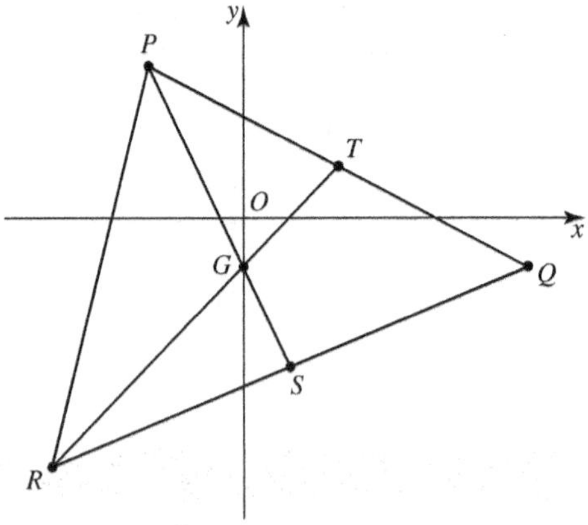

5 For triangle ABC with vertices $A(-5, 1)$, $B(7, 5)$ and $C(7, -9)$, show that the three medians are concurrent.

14 Applying algebraic skills to circles

Exercise 14.1 The circle equation from the centre and radius

1 Find the equations of the five circles shown in the diagram.

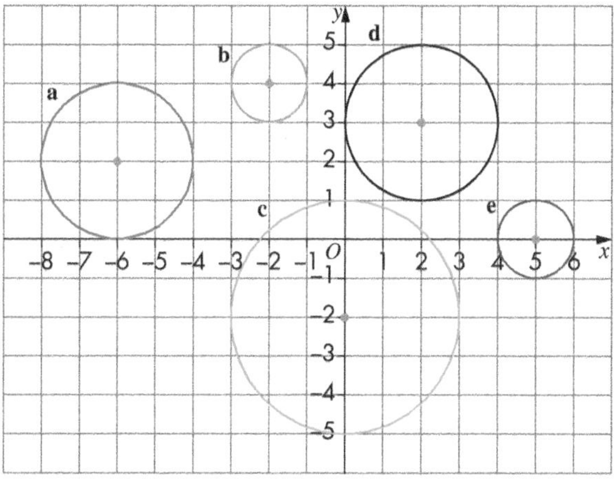

> **Hint** The equation of a circle with centre (a, b) and radius r is given by $(x-a)^2+(y-b)^2=r^2$

2 Find the equation of the circle with the given centre and radius.

a (2, 4); 6 **b** (–1, 3); 5 **c** (2, –3); 2 **d** (–3, –4); 1

e (0, –4); 2 **f** (–2, 0); 5 **g** (–3, 7); $\sqrt{3}$ **h** (10, –1); $\sqrt{17}$

3 State the centre and radius of these circles:

a $(x-1)^2+(y-2)^2=4$ **b** $(x+2)^2+(y-1)^2=9$ **c** $\left(x+\frac{1}{2}\right)^2+\left(y-\frac{3}{2}\right)^2=1$

d $(x-4)^2+\left(y+\frac{1}{4}\right)^2=3$ **e** $(x+10)^2+y^2=\frac{4}{9}$ **f** $(x+1)^2+\left(y+\frac{1}{3}\right)^2=36$

g $x^2+(y-1)^2=16$ **h** $x^2+y^2=\frac{1}{9}$ **i** $(x+6)^2+y^2=7$

4 Find the equations of the five circles shown in the diagram.

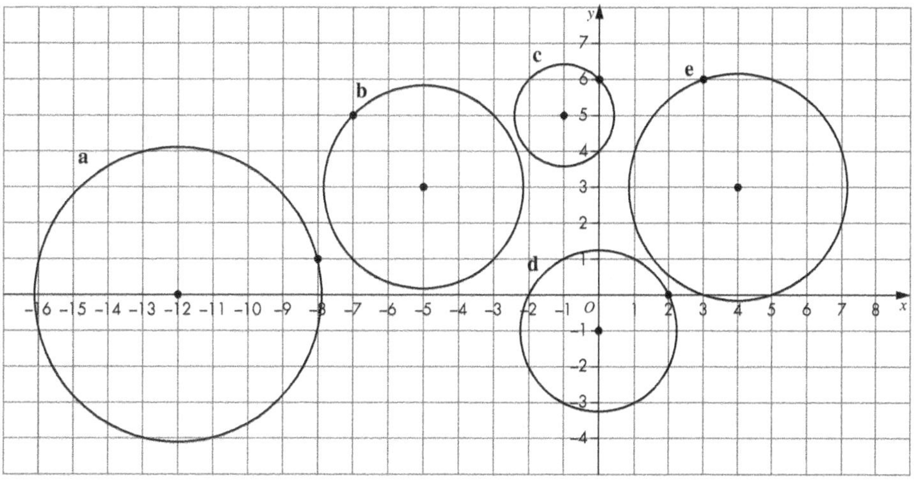

5 Find the equations of these circles where C is the centre and P lies on the circle:

a $C(-1, 2)$, $P(3, 5)$ **b** $C(4, -3)$, $P(-1, 9)$ **c** $C(0, 6)$, $P(2, 7)$

d $C(-2, 0)$, $P(1, -1)$ **e** $C(-3, -2)$, $P(0, 0)$ **f** $C(-15, 6)$, $P(-15, 0)$

g $C(-2, 3)$, $P(0, 3)$ **h** $C(\sqrt{2}, -\sqrt{2})$, $P(2\sqrt{2}, 2\sqrt{2})$

6 Find the equation of the circle that has P and Q as the endpoints of a diameter.

a $P(2, 3)$ and $Q(8, 1)$ b $P(-1, 2)$ and $Q(11, 18)$
c $P(2, 1)$ and $Q(-6, -5)$ d $P(-7, -2)$ and $Q(3, 18)$
e $P(0, -1)$ and $Q(2, -3)$ f $P(3, -8)$ and $Q(-1, 0)$
g $P\left(\frac{1}{2}, -\frac{1}{2}\right)$ and $Q\left(\frac{3}{2}, -\frac{3}{2}\right)$ h $P\left(\frac{1}{2}, -\frac{1}{2}\right)$ and $Q\left(-\frac{1}{2}, \frac{3}{2}\right)$

Hint The centre is the midpoint of the diameter.

Exercise 14.2 The general circle equation

1 Find the centre and radius of each circle.

a $x^2 + y^2 - 2x + 4y + 1 = 0$ b $x^2 + y^2 + 6x - 2y + 1 = 0$
c $x^2 + y^2 + 4x - 2y - 4 = 0$ d $x^2 + y^2 + 10x - 6y - 2 = 0$
e $x^2 + y^2 - 8x - 8y + 7 = 0$ f $x^2 + y^2 - x - y - \frac{1}{2} = 0$
g $x^2 + y^2 - 6x + 8y = 0$ h $2x^2 + 2y^2 - 20y - 22 = 0$

Hint For the equation $x^2 + y^2 + 2gx + 2fy + c = 0$ the centre is $(-g, -f)$ and the radius is $\sqrt{g^2 + f^2 - c}$

2 The following represent circles. Find the possible range of values of k.

a $x^2 + y^2 + 2x - 4y - k = 0$ b $x^2 + y^2 + 2y + k = 0$
c $x^2 + y^2 - 6x + k = 0$ d $x^2 + y^2 - x - y - k = 0$
e $x^2 + y^2 + 2kx - 8y + 25 = 0$ f $x^2 + y^2 + kx + ky + 2 = 0$

Hint The radius is positive.

Exercise 14.3 Tangents to circles

Example

Prove that $y = x + 1$ is a tangent to the circle with equation $x^2 + y^2 + 6x + 7 = 0$ and find the point of contact.

$x^2 + (x + 1)^2 + 6x + 7 = 0$ — Replace y with $x + 1$ in the circle equation.

$\Rightarrow x^2 + x^2 + 2x + 1 + 6x + 7 = 0$ — Multiply out and simplify.

$\Rightarrow 2x^2 + 8x + 8 = 0 \Rightarrow x^2 + 4x + 4 = 0$ — Divide both sides by 2 (common factor).

$\Rightarrow (x + 2)(x + 2) = 0 \Rightarrow x = -2$ — Factorise and solve the equation.

There is only one solution and so the line $y = x + 1$ is a tangent to the circle. — You must clearly state the reason for tangency.

When $x = -2$, $y = -2 + 1 = -1$ — Use the known value ($x = -2$) in the tangent equation.

The point of contact is $(-2, -1)$. — State the coordinates of the point of contact.

1 In each case show that the line is a tangent to the circle and find the point of contact.

a $y = 2 - x$ and $x^2 + y^2 = 2$
b $y = x - 1$ and $x^2 + y^2 + 2x - 1 = 0$
c $y + x + 1 = 0$ and $x^2 + y^2 - 4x + 2y + 3 = 0$
d $y - x = 5$ and $x^2 + y^2 + 6x + 4y + 5 = 0$

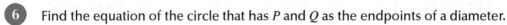

2 Match these circles with their tangents – there is one tangent to each circle.

Hint Draw sketches of the circles and lines. You are told they match.

Circles

Ⓐ $x^2 + (y - 2)^2 = 10$

Ⓑ $(x + 8)^2 + y^2 = 5$

Ⓒ $(x - 3)^2 + y^2 = 5$

Ⓓ $(x + 3)^2 + y^2 = 2$

Ⓔ $x^2 + (y + 2)^2 = 5$

Tangents

① $2y - x = 3$

② $y - 3x = -8$

③ $y - 2x = -1$

④ $y + x = -1$

⑤ $2y + x = -9$

3 Find the gradient of the tangent at P on the circle with centre C.

a $m_{CP} = \frac{2}{3}$ **b** $m_{CP} = 4$

c $m_{CP} = -\frac{1}{2}$ **d** $m_{CP} = -\frac{5}{4}$

e $m_{CP} = -1$ **f** $m_{CP} = 2$

g $m_{CP} = -6$ **h** $m_{CP} = 0$

Hint

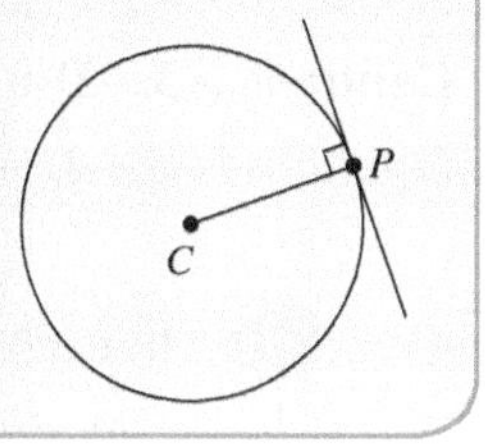

4 Find the equations of these circle tangents:

a

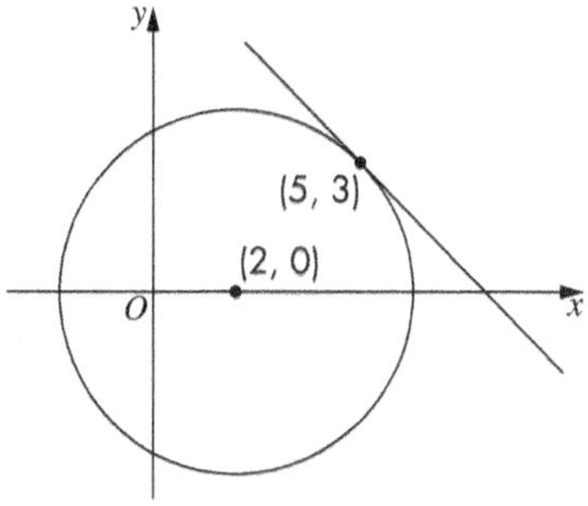

b

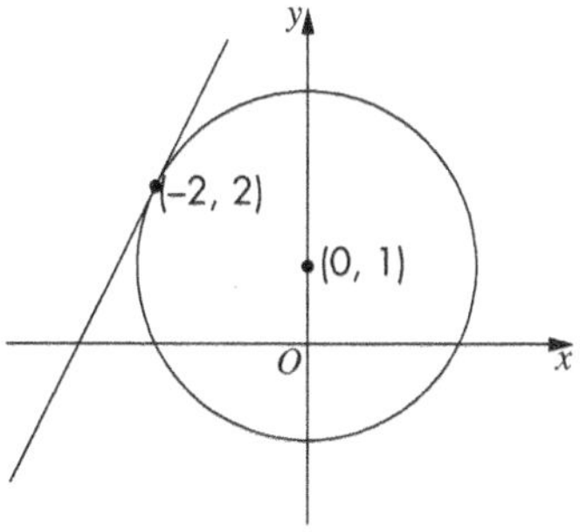

c

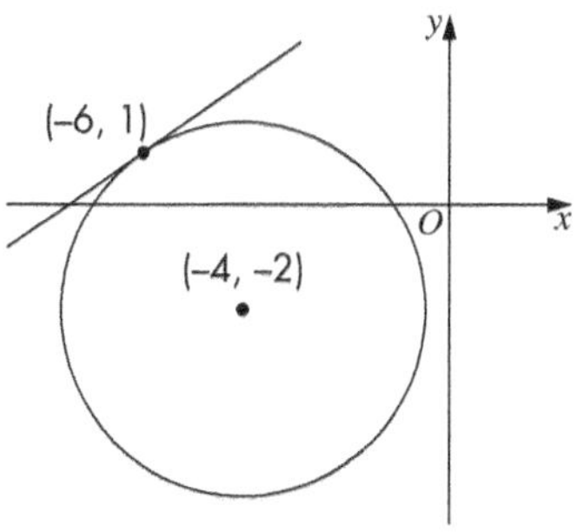

d

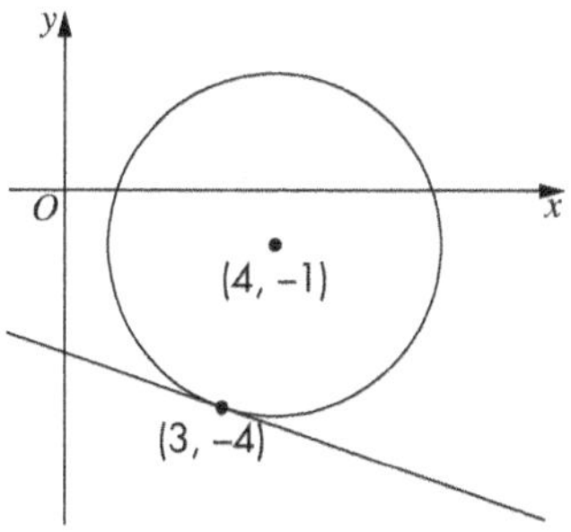

5 Find the equation of the tangent at the given point to the given circle.

a $(4, 3)$; $x^2 + y^2 - 4x - 2y - 3 = 0$ **b** $(-5, 5)$; $x^2 + y^2 + 6x - 2y - 10 = 0$

c $(0, -1)$; $x^2 + y^2 + 2x + 10y + 9 = 0$ **d** $(-5, 1)$; $x^2 + y^2 - 6y - 20 = 0$

e $(3, -4)$; $x^2 + y^2 - 16x + 35 = 0$ **f** $(-4, -3)$; $x^2 + y^2 = 25$

g $(-\frac{1}{2}, 1)$; $x^2 + y^2 - 5x - 6y + \frac{9}{4} = 0$ **h** $(5, 2)$; $x^2 + y^2 - 4y - 21 = 0$

Hint You should draw a sketch of each situation to help with your strategy and also to check that your answer makes sense.

Example

A circle passes through $A(-7, 0)$ and $B(1, 0)$ and has line $y = 2$ as a tangent.

Find the radius, centre and equation of the circle.

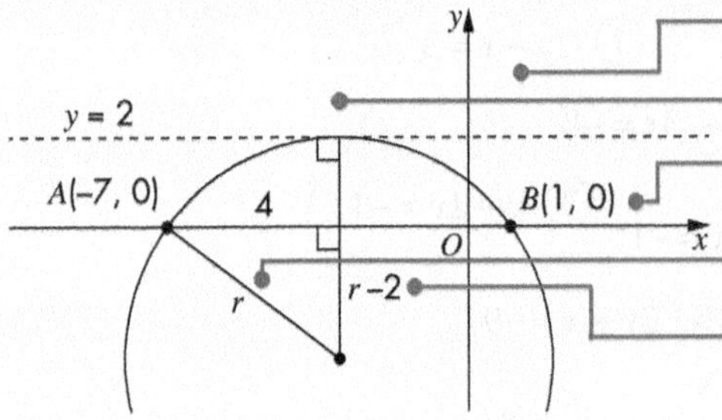

Draw a clear working sketch.

Note the right-angles since $y = 2$ is a tangent.

Fill in as many lengths and lines as needed.

Mark the unknown radius length as r.

The third side of the triangle has length $r - 2$.

$r^2 = 4^2 + (r - 2)^2$ — Use Pythagoras' theorem in the right-angled triangle.

$r^2 = 16 + r^2 - 4r + 4$

$4r = 20$ so $r = 5$

Centre is $(-3, -3)$ and radius is 5 — Work out the coordinates of the centre from the diagram.

Equation is $(x + 3)^2 + (y + 3)^2 = 25$

6 In each question use the information given to find the equation of the circle.

a A circle passes through the origin and the point $(30, 0)$ and has the line $y = -5$ as a tangent.

b The line $y = 2$ is a tangent to a circle which passes through $(0, 0)$ and $(12, 0)$.

c A circle which passes through the origin and has the line $x = -1$ as a tangent also passes through the point $(0, 10)$.

d A circle passing through $(-8, 0)$ and $(8, 0)$ has the line $y = 2$ as a tangent.

e The line $y = 1$ is a tangent to a circle which passes through $(-1, 0)$ and $(5, 0)$.

f A circle with $x = 1$ as a tangent passes through points $(0, 1)$ and $(0, 7)$.

Exercise 14.4 Intersecting and touching circles

1 Describe the relationship between these pairs of circles:

a $(x - 3)^2 + (y + 2)^2 = 4$ and $(x - 1)^2 + (y - 1)^2 = 1$

b $(x + 1)^2 + (y - 2)^2 = 4$ and $(x - 3)^2 + (y - 3)^2 = 5$

c $(x + 4)^2 + (y + 2)^2 = 5$ and $(x - 2)^2 + (y - 1)^2 = 20$

Hint Simplify surds, e.g. $\sqrt{45} = \sqrt{9 \times 5} = 3\sqrt{5}$

Hint

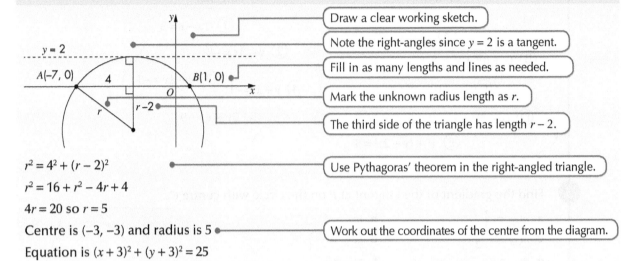

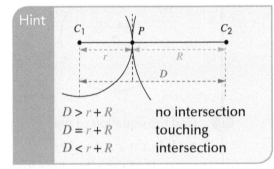

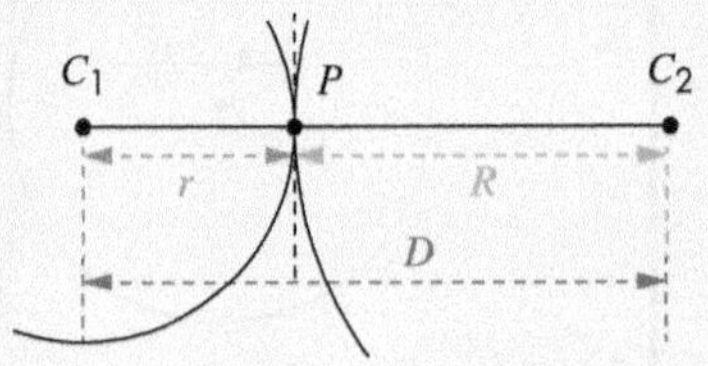

$D > r + R$	no intersection
$D = r + R$	touching
$D < r + R$	intersection

2 Determine whether or not the pair of equations represent circles that intersect.

a $(x + 2)^2 + y^2 = 4$ and $(x + 1)^2 + (y - 3)^2 = 9$

b $(x + 1)^2 + (y - 1)^2 = 2$ and $(x - 2)^2 + (y + 2)^2 = 8$

c $x^2 + y^2 + 4x + 2y = 0$ and $x^2 + y^2 - 12x - 6y = 0$

d $x^2 + y^2 + 6x - 4y + 3 = 0$ and $x^2 + y^2 - 12x + 2y - 3 = 0$

e $x^2 + y^2 + 8x - 14y + 52 = 0$ and $x^2 + y^2 - 10x - 2y - 26 = 0$

15 Modelling situations using sequences

Exercise 15.1 Sequences, nth term formulae and recurrence relations

1 For each of the following nth term formulae, find u_1, u_2, u_3, u_4 and u_{10}

a $u_n = n$ **b** $u_n = 2n - 1$ **c** $u_n = \frac{1}{n}$

d $u_n = 12 - n$ **e** $u_n = n^2$ **f** $u_n = 2$

g $u_n = 1 - \frac{1}{n}$ **h** $u_n = n^2 - n + 41$ **i** $u_n = (n + 1)(2n - 1)$

j $u_n = \frac{n}{n+1}$ **k** $u_n = -n + 1$ **l** $u_n = \frac{12}{n}$

m $u_n = 2^n$ **n** $u_n = (-1)^n$ **o** $u_n = 0{\cdot}1^n$

p $u_n = n^4 - 10n^3 + 35n^2 - 50n + 24$

> **Hint** To find u_k substitute k for n in the formula for u_n.

2 For each nth term formula, calculate u_5

a $u_n = 2^n$ **b** $u_n = 3n + 1$ **c** $u_n = 10 - 2n$

d $u_n = n^2 - 3$ **e** $u_n = 3^{n-2}$ **f** $u_n = 4n^2 - 5$

g $u_n = 2n + 1$ **h** $u_n = 2^n + 1$ **i** $u_n = 2^{n+1}$

3 Use each recurrence relation and u_1 value to calculate the next four terms of the sequence.

a $u_{n+1} = u_n + 2$ and $u_1 = 5$ **b** $u_{n+1} = u_n - 2$ and $u_1 = 10$

c $u_{n+1} = u_n + 0{\cdot}1$ and $u_1 = 0$ **d** $u_{n+1} = u_n - 1$ and $u_1 = -4$

e $u_{n+1} = 2u_n$ and $u_1 = 1$ **f** $u_{n+1} = 3u_n$ and $u_1 = 1$

g $u_{n+1} = 0{\cdot}1u_n$ and $u_1 = 3$ **h** $u_{n+1} = 2u_n - 2$ and $u_1 = 2$

i $u_{n+1} = 2u_n - 2$ and $u_1 = 3$ **j** $u_{n+1} = 2u_n - 2$ and $u_1 = 1$

> **Hint** Build the sequence from u_1 using the recurrence relation to calculate first u_2 then u_3 then u_4 then u_5.

4 For each recurrence relation calculate u_5

a $u_{n+1} = 2u_n$ with $u_1 = 1$ **b** $u_{n+1} = \frac{1}{2}u_n + 4$ with $u_1 = 24$

c $u_{n+1} = 12 - 2u_n$ with $u_1 = 3$ **d** $u_{n+1} = \frac{1}{2}u_n$ with $u_1 = 256$

e $u_{n+1} = 3u_n - 1$ with $u_1 = 1$ **f** $u_{n+1} = u_n + 10$ with $u_1 = 2$

5 For each of these sequences, write down a recurrence relation, using u_n and u_{n+1}, that the terms of the sequence satisfy:

a 2, 3, 4, 5, … **b** 1, 6, 11, 16, 21, …

c 25, 23, 21, 19, … **d** 1, 2, 4, 8, 16, …

e 1, 3, 9, 27, 81, … **f** 1, 10, 100, 1000, …

g 1, 0·1, 0·01, 0·001, … **h** 8, 4, 0, −4, …

Example

The sequence 3, 12, 18, 22, ... is generated by a recurrence relation of the form $u_{n+1} = au_n + b$.
Find the values of a and b.

$12 = a \times 3 + b \Rightarrow 3a + b = 12$ — $u_1 = 3$ and $u_2 = 12$, now use $u_2 = a \times u_1 + b$

$18 = a \times 12 + b \Rightarrow 12a + b = 18$ — $u_2 = 12$ and $u_3 = 18$, now use $u_3 = a \times u_2 + b$

Solve:
$$\left.\begin{aligned} 3a + b &= 12 \\ 12a + b &= 18 \end{aligned}\right\}$$
Solve these equations simultaneously.

Subtract top equation from bottom equation.

$\Rightarrow 9a = 6 \Rightarrow a = \frac{2}{3}$

Substitute $a = \frac{2}{3}$ in $3a + b = 12$.

So $3 \times \frac{2}{3} + b = 12 \Rightarrow 2 + b = 12 \Rightarrow b = 10$

Check: $u_{n+1} = \frac{2}{3}u_n + 10$ — Start your checking with $u_1 = 3$: $\frac{2}{3} \times 3 + 10 = 12$, then check $\frac{2}{3} \times 12 + 10 = 18$, etc.

6 Each of these sequences is generated using a recurrence relation of the form $u_{n+1} = au_n + b$
In each case find the values of a and b.

a 3, 5, 9, 17, ... **b** 2, 8, 26, 80, ... **c** 2, 6, 8, 9, ... **d** 15, 6, 3, 2, ...
e 11, 7, 5, 4, ... **f** 2, 6, 26, 126, ... **g** 1, −1, 3, −5, ... **h** $\sqrt{2}, 1, \sqrt{2} - 1, 1 - \sqrt{2}, \ldots$

Exercise 15.2 Sequences and limits

1 For the sequence defined by each recurrence relation explain whether or not there is a limit as n tends to infinity. If there is a limit then find its **exact** value.

a $u_{n+1} = 0{\cdot}4u_n + 1$ **b** $u_{n+1} = 0{\cdot}7u_n + 8$ **c** $2u_{n+1} = 3u_n + 8$
d $u_{n+1} = \frac{u_n}{4} + 5$ **e** $7u_{n+1} = 2u_n + 8$ **f** $u_{n+1} = \frac{1}{3}u_n - 1$
g $11(u_{n+1} - 1) = 4u_n$ **h** $23u_{n+1} - 14u_n = 46$

Hint For a sequence generated by $u_{n+1} = mu_n + c$ if $-1 < m < 1$ there is a limit L where $L = \frac{c}{1 - m}$

2 Birds eat 5% of the seeds stored in a barn each day so the farmer then tops them up with a further 3 kg of seeds. Initially there were 100 kg of seeds. In the long term, what will be the weight of the seeds?

3 60% of pollutant is removed each day by a water-purifying machine but 10 litres of new pollutant is then added. If the water initially contained 20 litres of pollutant, what will be the long-term level of the pollutant?

4 15% of an ant colony dies each day but 150 new ants are born each night. If the colony starts with 500 ants what is its final size?

5 The body destroys 70% of a drug in a day so a daily injection of 28 units is given. The initial injection was 50 units.

a At no time should the body contain more than 50 units of the drug. Is this course of treatment safe?

b Is it safe to increase the daily injections to 36 units?

6 The air pressure in a bouncy castle reduces by 12% from a day's use so the owner increases the pressure by 9 units for the next day. Initially the pressure was 70 units. It is dangerous to operate at greater than 74·5 units pressure. Is what the owner is doing safe?

16 Applying differential calculus

Exercise 16.1 Maxima and minima on intervals

For Questions 1 and 2, use the information given in each diagram of the graph of $y = f(x)$ to find the minimum and maximum value taken by the function f on the various intervals.

> **Hint** Maximum and minimum values are found only at stationary points or at the endpoints of the given interval.

1 **a** **i** $-5 \leqslant x \leqslant 1$
ii $-6 \leqslant x \leqslant 1$
iii $-2 \leqslant x \leqslant 2$
iv $-3 \leqslant x \leqslant 0$
v $-3 \leqslant x \leqslant -1$

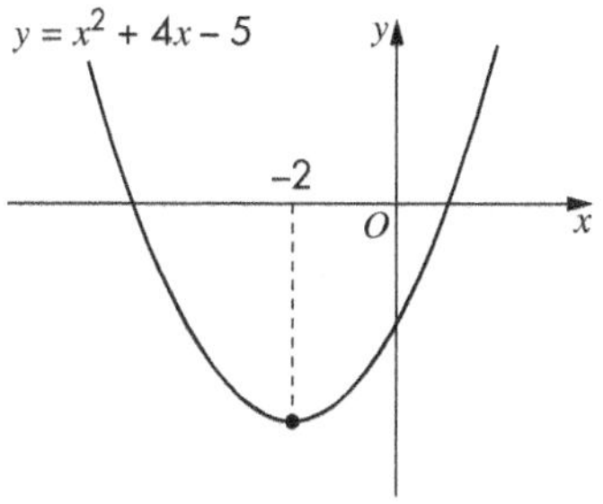

b **i** $0 \leqslant x \leqslant 4$
ii $1 \leqslant x \leqslant 5$
iii $3 \leqslant x \leqslant 8$
iv $-1 \leqslant x \leqslant 5$
v $-2 \leqslant x \leqslant 9$

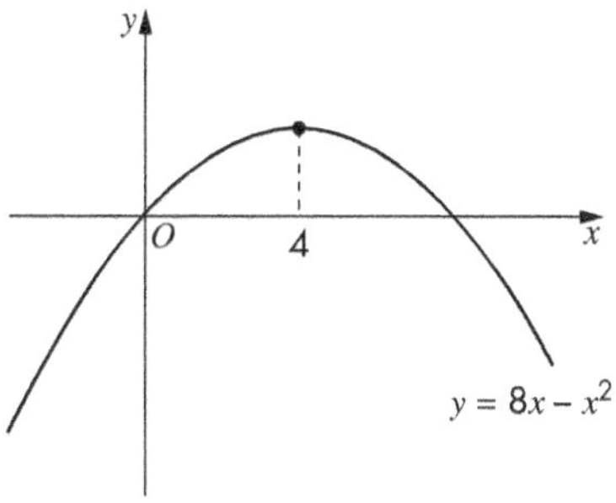

c **i** $-4 \leqslant x \leqslant 0$
ii $2 \leqslant x \leqslant 3$
iii $-5 \leqslant x \leqslant 1$
iv $-1 \leqslant x \leqslant 1$
v $-7 \leqslant x \leqslant 1$

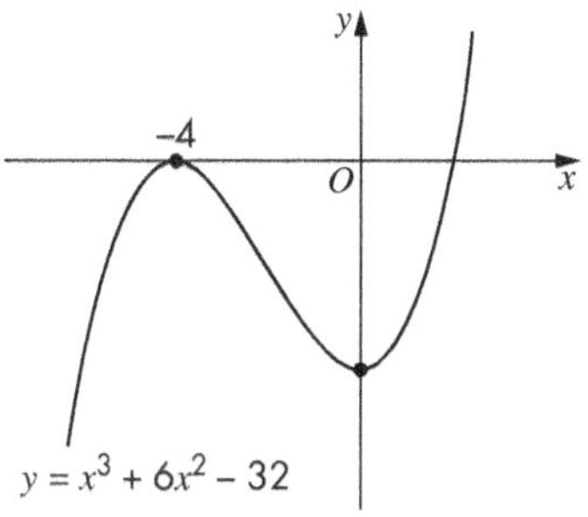

d **i** $-1 \leqslant x \leqslant 0$
ii $0 \leqslant x \leqslant 1$
iii $-2 \leqslant x \leqslant 0$
iv $0 \leqslant x \leqslant 2$
v $-2 \leqslant x \leqslant 2$

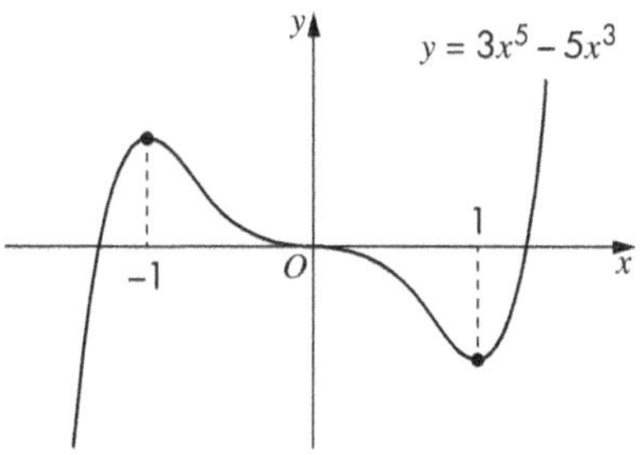

2 a i $0 \leqslant x \leqslant \frac{\pi}{2}$

ii $\frac{\pi}{2} \leqslant x \leqslant \frac{3\pi}{2}$

iii $0 \leqslant x \leqslant 2\pi$

iv $\frac{\pi}{3} \leqslant x \leqslant \frac{4\pi}{3}$

v $-12\pi \leqslant x \leqslant 12\pi$

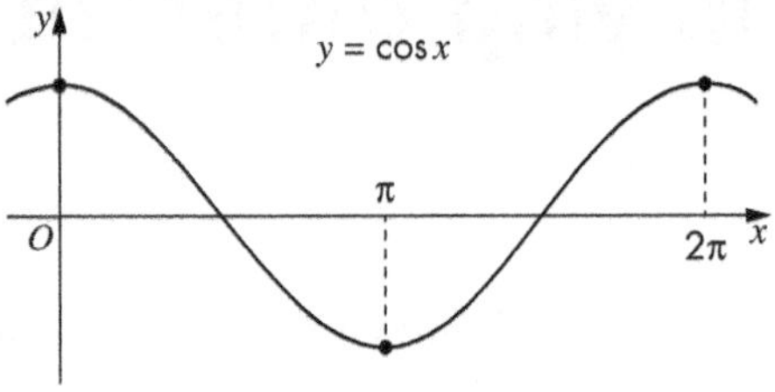

b i $0 \leqslant x \leqslant 2\pi$

ii $\pi \leqslant x \leqslant \frac{3\pi}{2}$

iii $\frac{\pi}{4} \leqslant x \leqslant \frac{3\pi}{4}$

iv $-\frac{\pi}{4} \leqslant x \leqslant \frac{\pi}{4}$

v $\frac{\pi}{4} \leqslant x \leqslant \frac{5\pi}{4}$

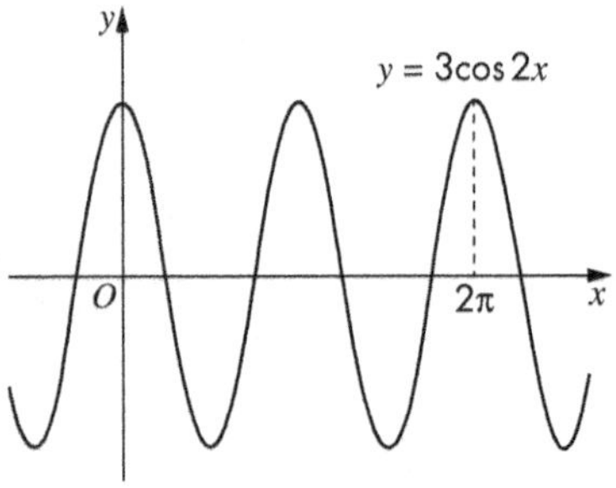

c i $0 \leqslant x \leqslant \frac{\pi}{6}$

ii $0 \leqslant x \leqslant \frac{7\pi}{6}$

iii $0 \leqslant x \leqslant 2\pi$

iv $\frac{\pi}{3} \leqslant x \leqslant \pi$

v $\frac{\pi}{6} \leqslant x \leqslant \frac{4\pi}{3}$

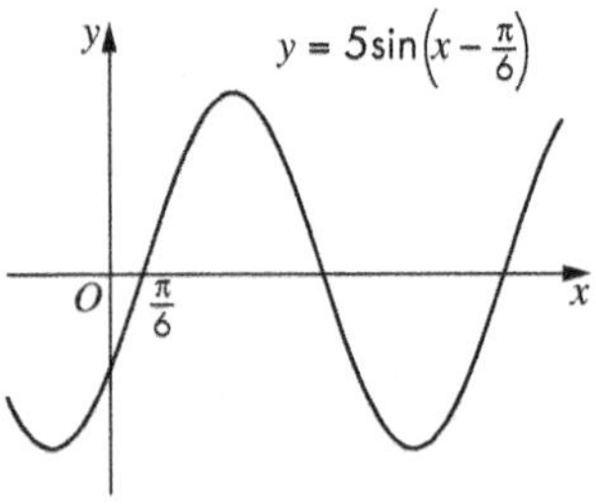

Example

Identify the stationary points and their nature for the graph of $y = f(x)$ where $f(x) = x^3 - 6x^2$ and hence determine the maximum and minimum values of f on the interval $-1 \leqslant x \leqslant 3$.

$f(x) = x^3 - 6x^2 = x^2(x - 6)$

$f'(x) = 3x^2 - 12x = 3x(x - 4)$

These factorised forms are easier for calculations.

$f'(x) = 0 \Rightarrow x = 0$ or $x = 4$

Solve $f'(x) = 0$ to find where the stationary points are.

x	0^-	0	0^+
$f'(x)$	+	0	−
Shape	/	—	\

x	4^-	4	4^+
$f'(x)$	−	0	+
Shape	\	—	/

These tables determine the nature of the two stationary points.

$f(0) = 0^2 \times (0 - 6) = 0 \Rightarrow (0, 0)$ is a max S.P.

$f(4) = 4^2 \times (4 - 6) = -32 \Rightarrow (4, -32)$ is a min S.P.

Calculate the y-coordinates of the stationary points.

$f(-1) = (-1)^2 \times (-1 - 6) = -7 \Rightarrow (-1, -7)$ endpoint

$f(3) = 3^2 \times (3 - 6) = -27 \Rightarrow (3, -27)$ endpoint

Calculate the y-coordinates of the endpoints of the interval.

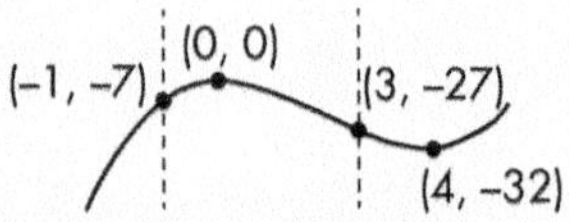

A sketch helps to understand the situation.

Max value is 0 and min value is −27 on $-1 \leqslant x \leqslant 3$.

Final statement of values.

3 In each case identify the stationary points and their nature for the graph of $y = f(x)$ and hence determine the maximum and minimum values of f on the given interval.

a $f(x) = x^2 - 4x - 12$ on the interval $-3 \leqslant x \leqslant 0$.

b $f(x) = 4x - x^2$ on the interval $-1 \leqslant x \leqslant 1$.

c $f(x) = -x^2 + 6x - 5$ on the interval $0 \leqslant x \leqslant 5$.

d $f(x) = -\frac{1}{3}x^3$ on the interval $-1 \leqslant x \leqslant 1$.

e $f(x) = x^3 - 3x^2$ on the interval $-1 \leqslant x \leqslant 4$.

f $f(x) = 6x^2 - x^3$ on the interval $-2 \leqslant x \leqslant 5$.

g $f(x) = 3x^2 - x^3$ on the interval $-\frac{1}{2} \leqslant x \leqslant \frac{5}{2}$.

h $f(x) = 9x^2 - \frac{1}{2}x^4$ on the interval $-3 \leqslant x \leqslant 2$.

> **Hint** Stationary points can be found by solving the equation $f'(x) = 0$

4 For each situation, give the interval of valid values for the variable x.

a A rectangular card

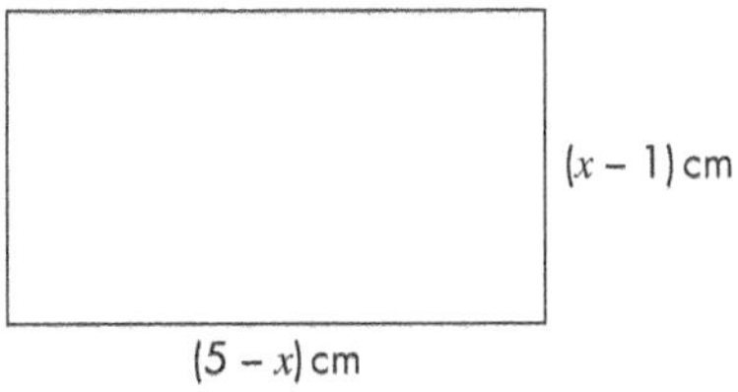

b A cardboard box

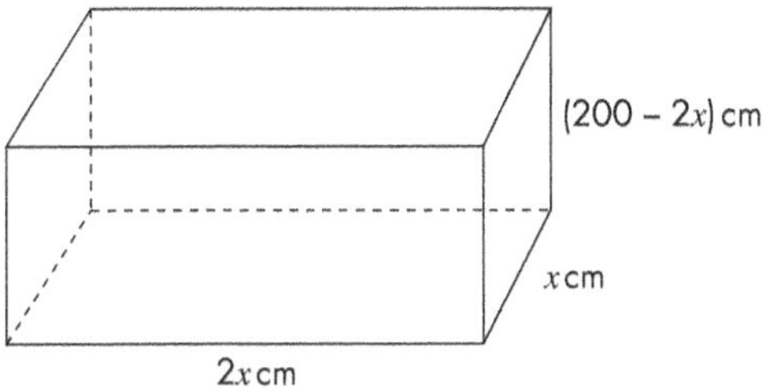

c A rectangular shaped post-clip

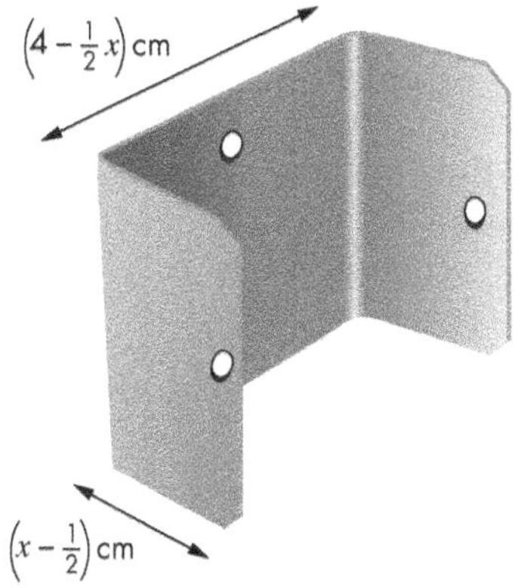

d A cylinder

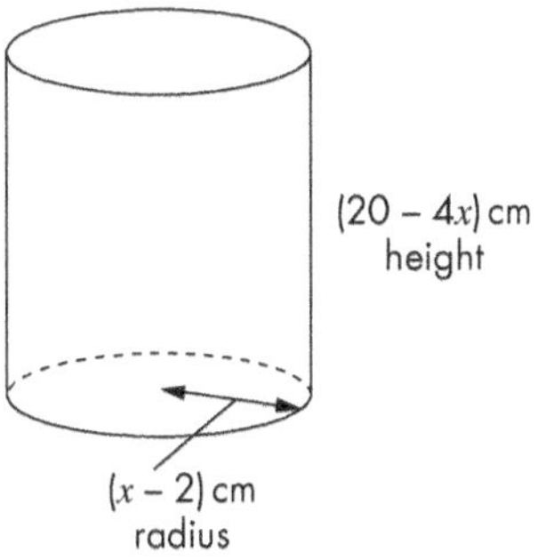

e A 20 cm × 20 cm sheet of paper with four identical squares cut from the corners

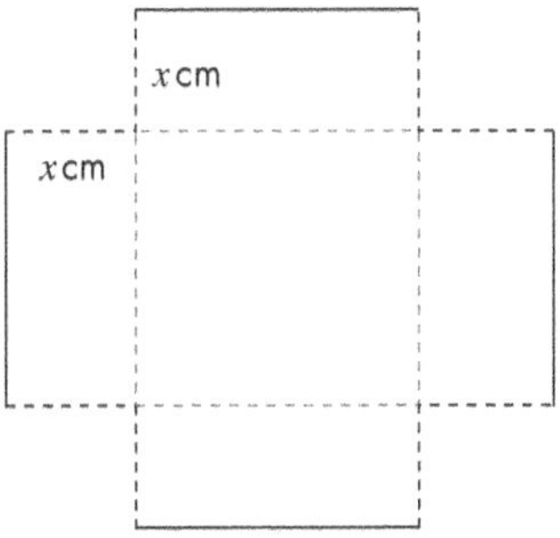

f A 2 m² rectangular poster with a fixed-width internal 0·2 metre wide border

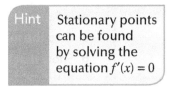

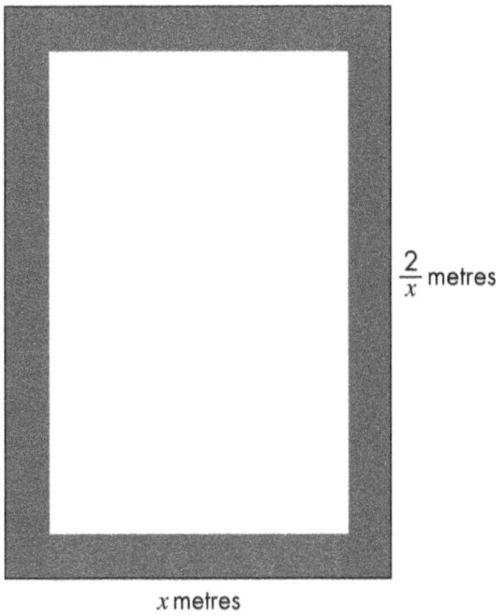

Exercise 16.2 Problems involving optimisation

1 The diagram shows the design for a wooden wall plaque.

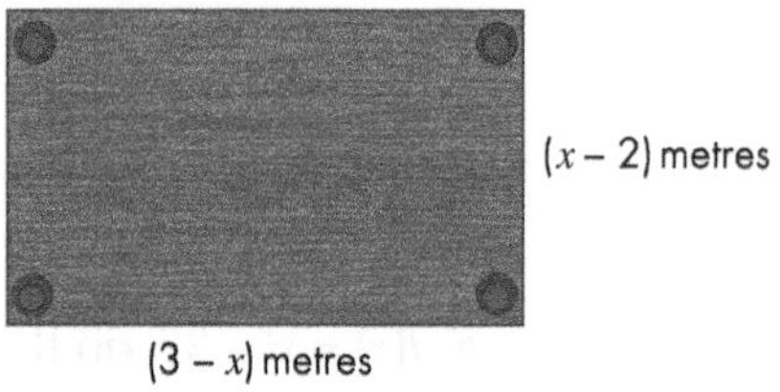

a Give the interval of valid values for x.

b Show that the area, A m^2, is given by $A(x) = -x^2 + 5x - 6$

c Find the value of x that gives the maximum area for the plaque.

d Find the dimensions of the plaque that give this maximum area.

e Find the maximum area possible for this plaque.

2 The cost, £C million, of manufacturing x thousand cars in a one-year period at a particular factory is given by:

$$C(x) = x^3 - \frac{45}{2}x^2 + 132x$$

For practical reasons $1 \leqslant x \leqslant 14$.

a Find the cost of manufacturing 10 000 cars at this factory in a one-year period.

b Calculate C at the endpoints of the given interval.

c Find the stationary values of C and determine their nature.

d What should the factory do to minimise its production costs?

3 An airline is investigating the total profit, £P, they make for each flight on a particular route. If there are n passengers on average on each flight they discovered that: $P(n) = 800n - \frac{5}{3}n^2$ when the ticket price was £$(800 - \frac{5}{3}n)$

a If the maximum capacity for each plane is 300 and it is uneconomical to run a flight with fewer than 180 passengers, give the interval of realistic values for n.

b Calculate $P(180)$ and $P(300)$.

c Find the stationary value for P and determine its nature.

d For the airline to maximise its profits, how many passengers should they aim for on each flight and what should they charge them?

4 **a** A metal can has a volume of 330 ml.

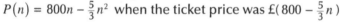

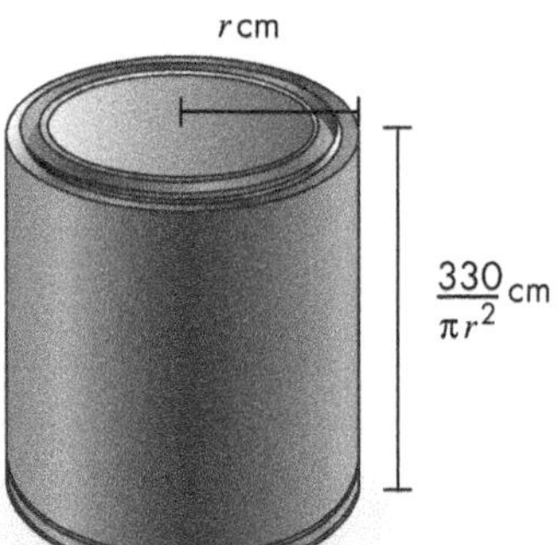

The surface area, A cm^2, of the can is given by:

$A(r) = 2\pi r^2 + \frac{660}{r}$, where the radius of the base is r cm.

Show that there is a stationary value for A when $r = \sqrt[3]{\frac{660}{2\pi}}$

b Given that this is a minimum stationary value for A, find the dimensions of the can (diameter and height) that uses the least amount of metal to produce. Give your answer correct to 2 decimal places.

5 The sum of two positive numbers is 120. P is the product of one of these numbers and the square of the other.

a If one of the numbers is x, write down an expression for the other number.

b Show that $P(x) = 120x^2 - x^3$.

c Give the interval of allowed values for x.

d Find the stationary values for P and determine their nature.

e What are the two numbers that give the greatest product P?

Hint: The product of x and y is xy.

6 Two ships leave two different ports at noon. After t hours they are D km apart where $D(t) = 9t^2 + (13 - 2t)^2$.

a How far apart are the two ports?

b Are there any restrictions on the values of t?

c Find the stationary value of D and determine its nature.

d What is the closest the two ships get and when does this happen?

Hint: The ships leave port when $t = 0$.

7 An international haulage company is trying to minimise their costs. When their lorries travel at v km/h the fuel used costs £C per hour where

$$C(v) = \frac{v}{4} + \frac{1600}{v}$$

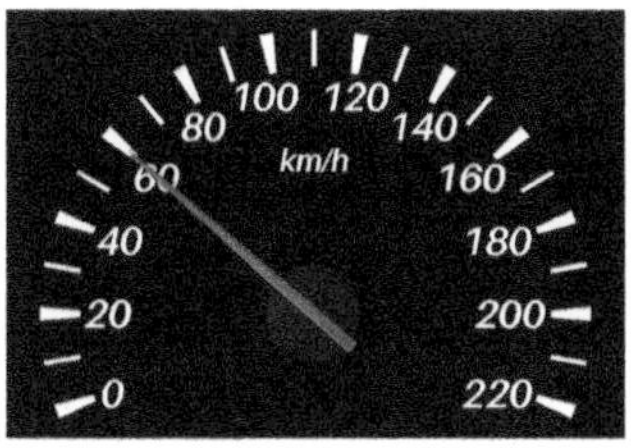

Lorries have a maximum speed limit of 100km/h.

a Give the interval of allowed values for v.

b Find the stationary values for C and determine their nature.

c What advice should the company give their drivers in order to minimise their fuel costs?

8 A design for a decorative metal storage bin is shown in the diagram.

It has a square base of side x metres and has a volume of 13·5m^3.

a Show that $h = \frac{27}{2x^2}$ m

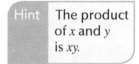

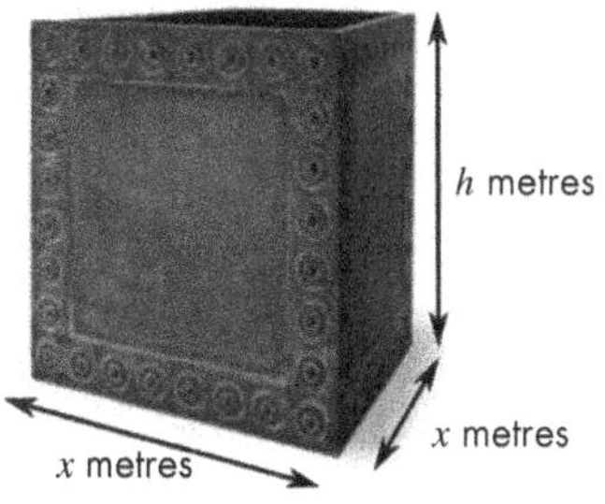

b Show that the area of each of the four sides is $\frac{27}{2x}$ m^2

c The bin has no top. Show that the total surface area, A m^2, is given by:

$$A(x) = x^2 + \frac{54}{x}$$

d Find the stationary points of A and determine their nature.

e What dimensions give the most economical (least surface area) cost of making these bins?

Exercise 16.3 Problems involving rates of change

Reminder

s is displacement v is velocity a is acceleration t is time

The rate of change of displacement gives velocity: $\frac{ds}{dt} = v$

The rate of change of velocity gives acceleration: $\frac{dv}{dt} = a$

1. The displacement, s cm, of a weight on a spring t seconds after release is given by: $s = 50t - 100t^2$

   Find its velocity:

   a $\frac{1}{10}$ second after release

   b $\frac{1}{4}$ second after release. Explain your answer.

2. Three stones are thrown up into the air at the same time. The height, h metres, of the stones t seconds after they are thrown up is given by these formulae:

   Stone 1: $h = 60t - 5t^2$ Stone 2: $h = 56t - 7t^2$ Stone 3: $h = 16t - 4t^2$

   For each stone find its velocity:

   a when it is thrown

   b 2 seconds after it is thrown

   c 4 seconds after it is thrown.

3. A miniature rocket is launched. The height, h metres, of the rocket t minutes after launch is given by $h = 400t - 200t^2$.

   a Find the velocity of the rocket at launch.

   b Find the velocity of the rocket after 1 minute. Explain your answer.

   c Compare the velocity of the rocket after 2 minutes with its velocity at launch. Explain your result.

4. The height, h metres, of a football t seconds after being kicked is given by:

   $h(t) = -4{\cdot}9t^2 + 4{\cdot}9t + 1{\cdot}5$

   a Find an expression for the velocity, v m/s, in terms of t.

   b Find the velocity of the football after:

   i $\frac{1}{10}$ second **ii** $\frac{1}{4}$ second

   c After how long is the velocity zero and at what height?

5. A ball is thrown downwards from a drone. Its displacement, s feet from the ground, after t seconds is given by: $s(t) = -16t^2 - 64t + 512$

   a How high was the drone?

   b How long does the ball take to hit the ground?

   c What was the ball's velocity when it hit the ground?

   d What can you say about the ball's acceleration? Explain your answer.

17 Applying integral calculus

Exercise 17.1 Area between an algebraic graph and the x-axis

1 For each diagram:

i find the length and breadth of the rectangle and hence calculate its area

ii evaluate the given definite integral. Does it give the same result?

a

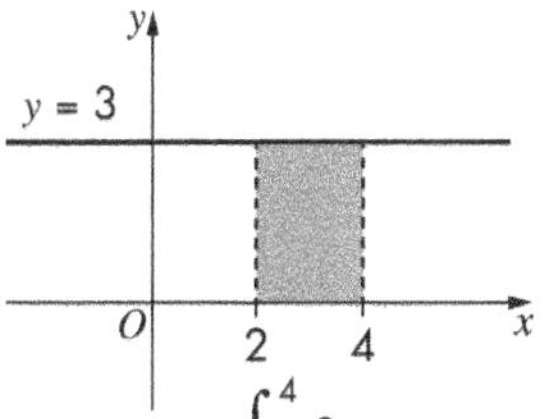

$\int_{2}^{4} 3\,dx$

b

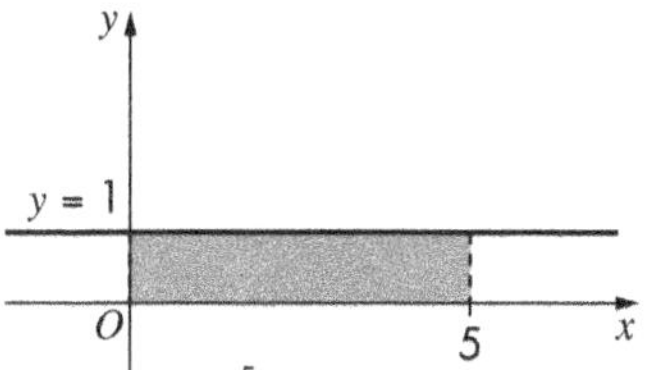

$\int_{0}^{5} 1\,dx$

c

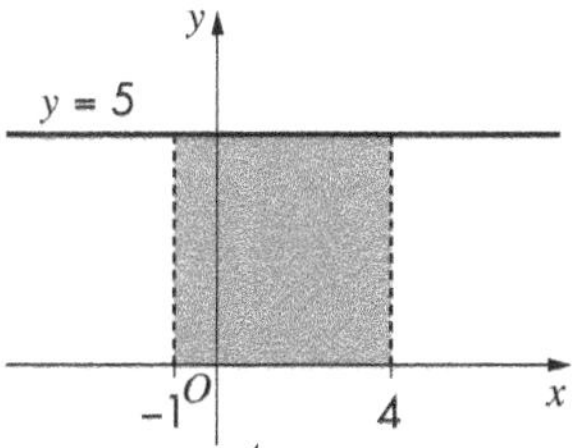

$\int_{-1}^{4} 5\,dx$

d

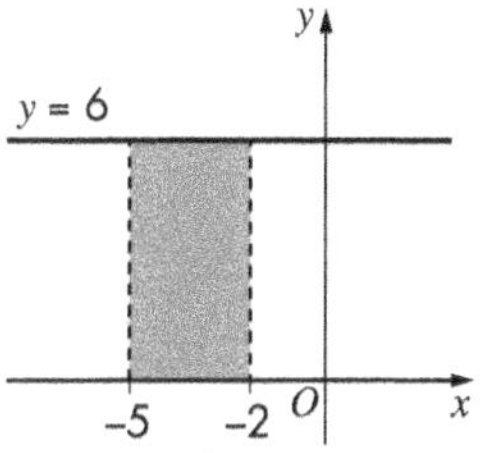

$\int_{-5}^{-2} 6\,dx$

2 For each diagram:

i use the squares on the grid to calculate the area of the shaded shape

ii show that the given definite integral gives the same result.

a

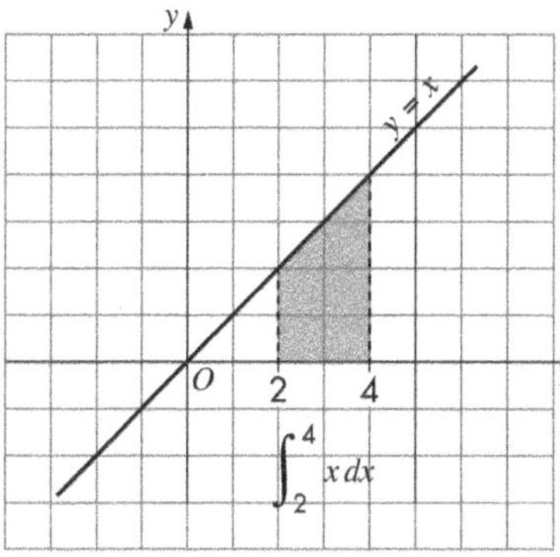

$\int_{2}^{4} x\,dx$

b

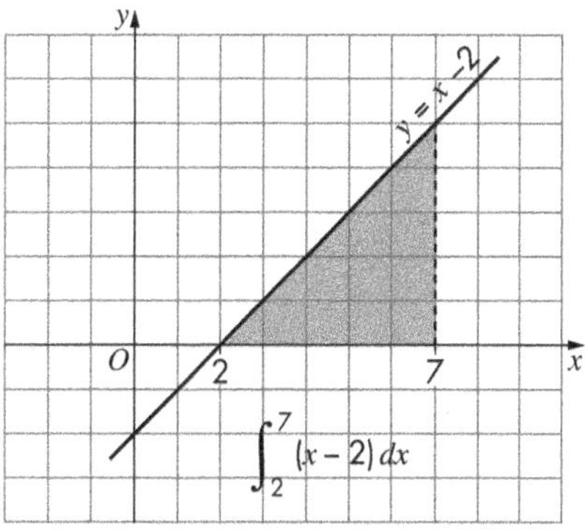

$\int_{2}^{7} (x-2)\,dx$

c

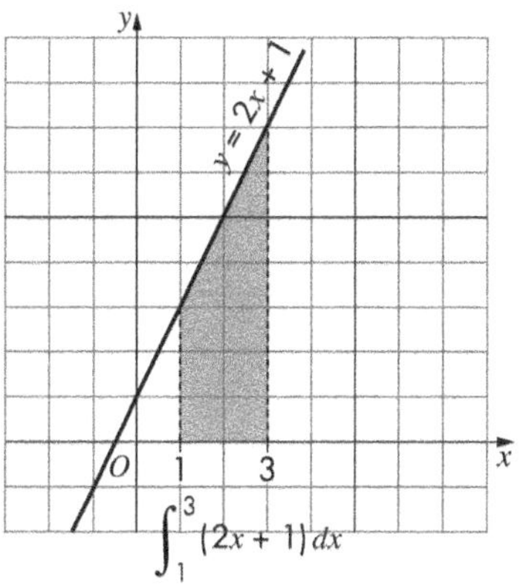

$\int_{1}^{3} (2x+1)\,dx$

d

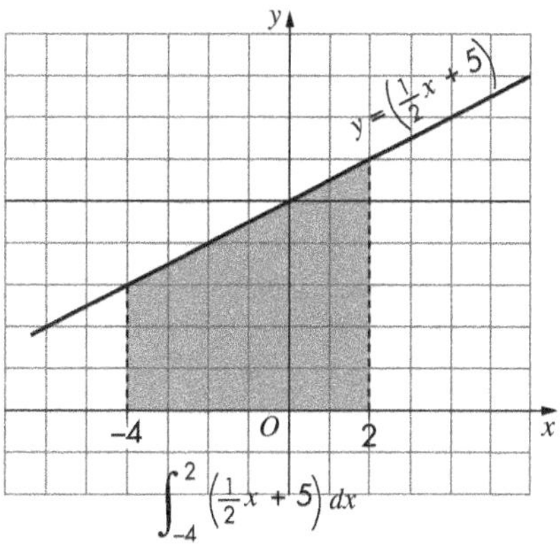

$\int_{-4}^{2} \left(\frac{1}{2}x+5\right)dx$

3   For each diagram:
    i   count grid squares to calculate Area I, Area II and the total shaded area
    ii evaluate the three definite integrals.
    iii what is the relationship between the three integrals?
    iv Explain which integrals are used to find the total shaded area.

a

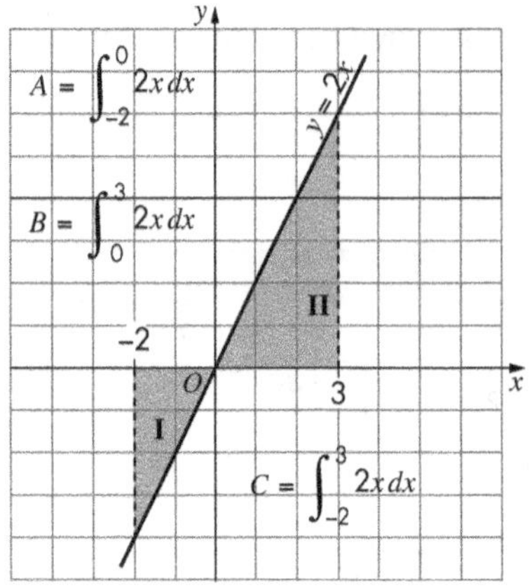

$$A = \int_{-2}^{0} 2x\,dx$$

$$B = \int_{0}^{3} 2x\,dx$$

$$C = \int_{-2}^{3} 2x\,dx$$

b

$$A = \int_{-6}^{0} x\,dx$$

$$B = \int_{0}^{4} x\,dx$$

$$C = \int_{-6}^{4} x\,dx$$

c

$$A = \int_{0}^{3} (3-x)\,dx$$

$$B = \int_{0}^{7} (3-x)\,dx$$

$$C = \int_{3}^{7} (3-x)\,dx$$

d

$$A = \int_{-2}^{2} \left(1 - \tfrac{1}{2}x\right) dx$$

$$B = \int_{2}^{6} \left(1 - \tfrac{1}{2}x\right) dx$$

$$C = \int_{-2}^{6} \left(1 - \tfrac{1}{2}x\right) dx$$

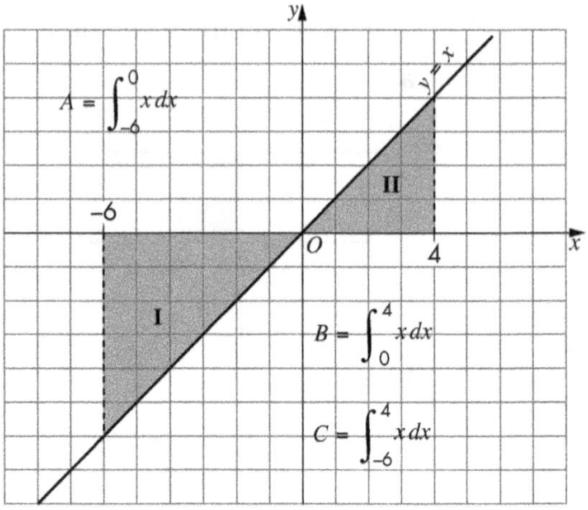

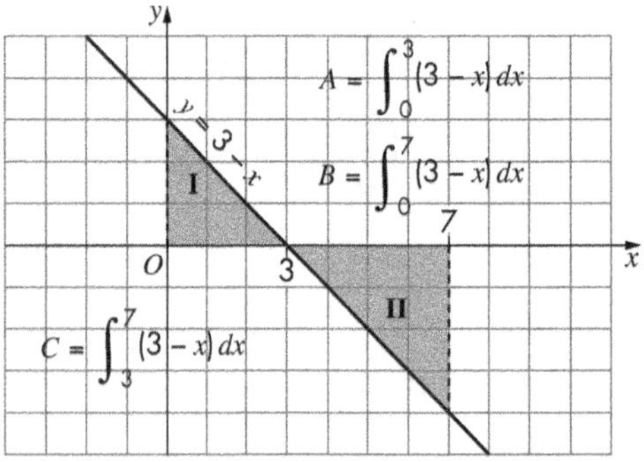

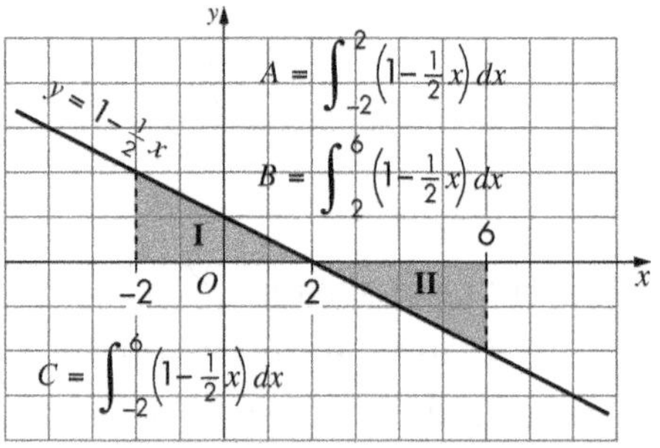

4   Calculate the exact shaded area in each of the following.

a

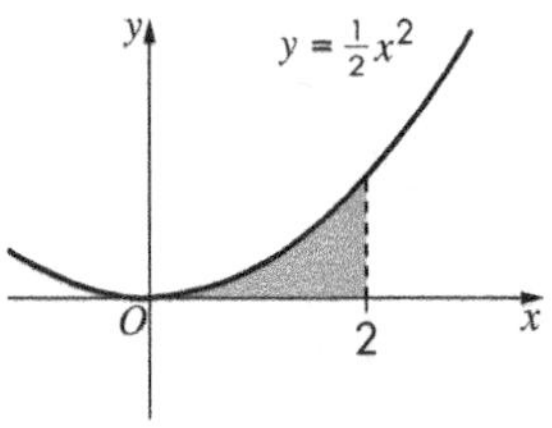

$y = \tfrac{1}{2}x^2$

b

$y = \tfrac{1}{2}x^2$

Hint

$y = f(x)$

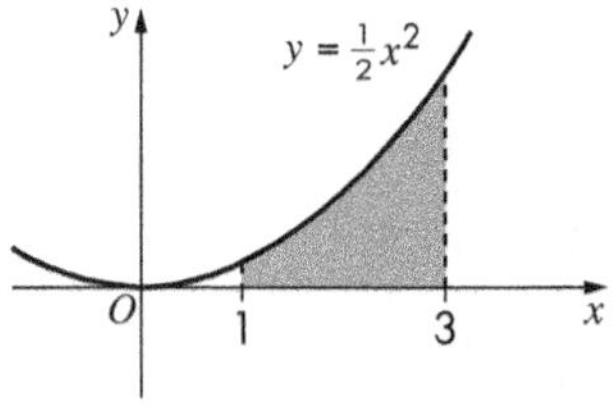

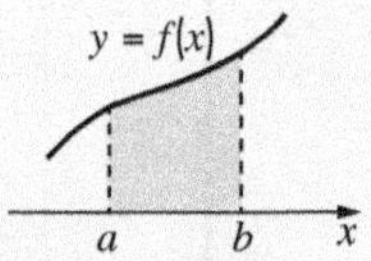

The shaded area is given by the integral

$$\int_{a}^{b} f(x)\,dx$$

c

$y = x^2$

d

$y = x^2$

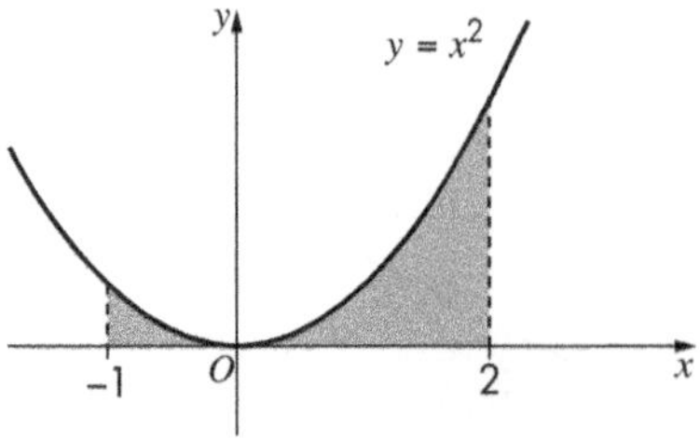

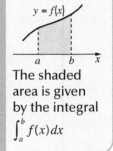

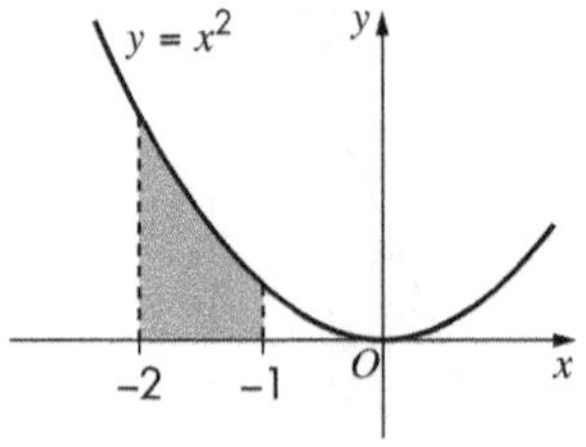

5 Each shaded area in these diagrams is exactly $\frac{4}{3}$ units². For each area, construct the appropriate definite integral and then show the calculations necessary to give that area.

a

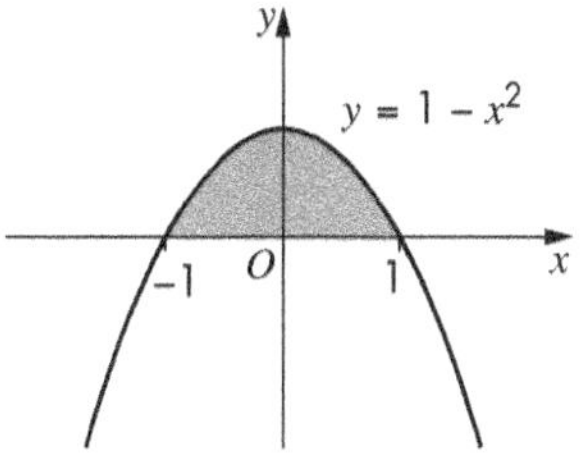

b

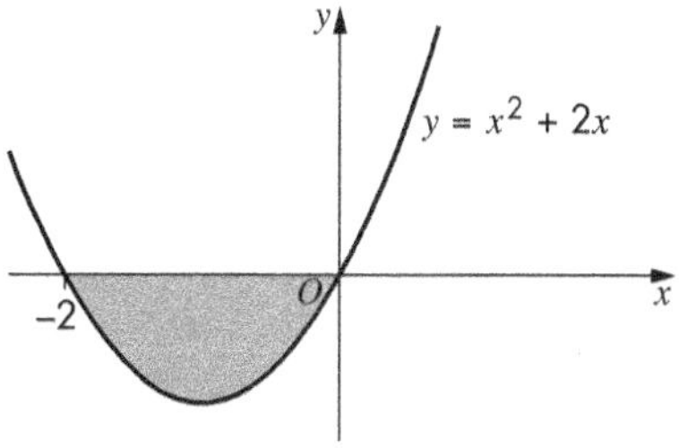

> **Hint** The definite integral will give a negative value for areas below the x-axis.

c

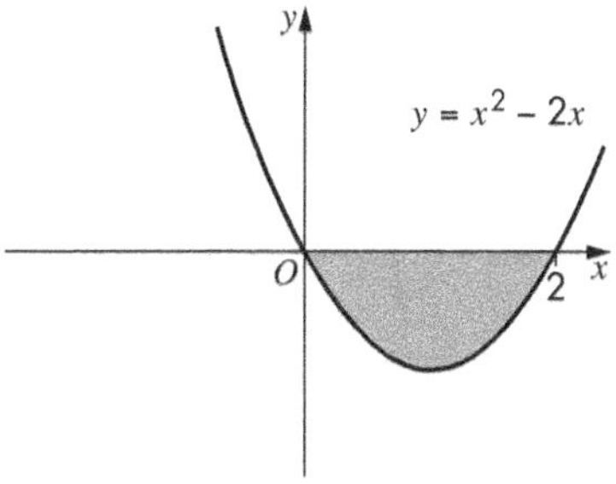

d

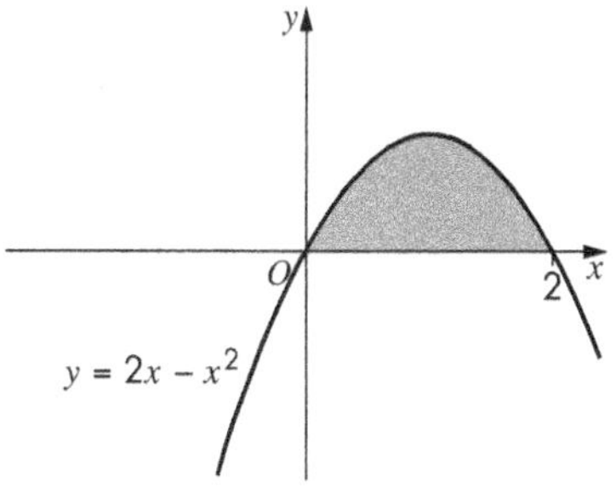

e

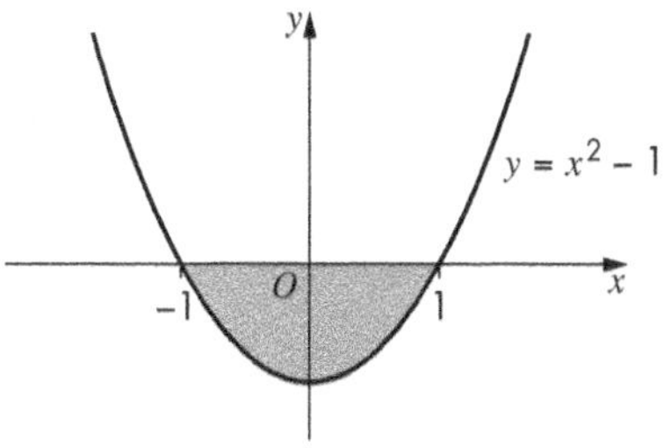

f

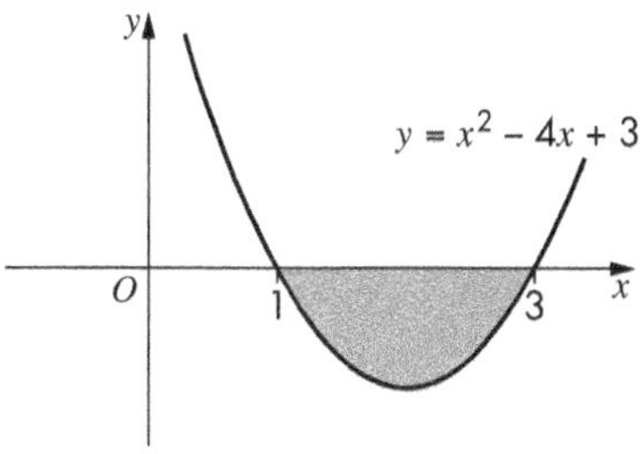

6 Calculate the total shaded area in each diagram.

a

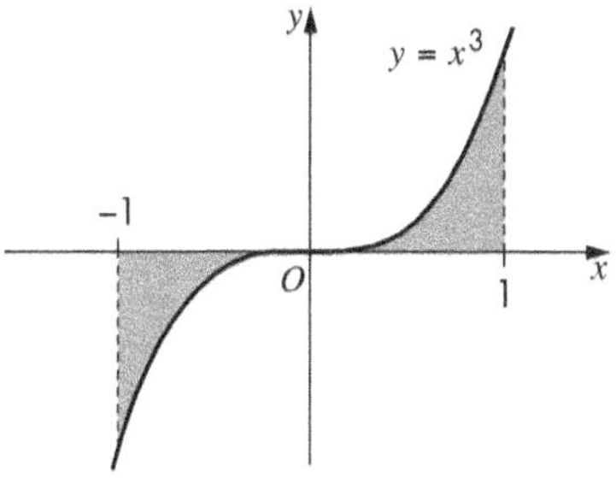

b

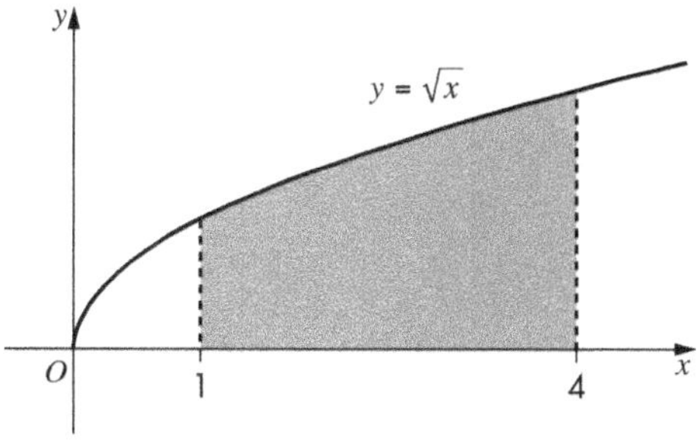

c

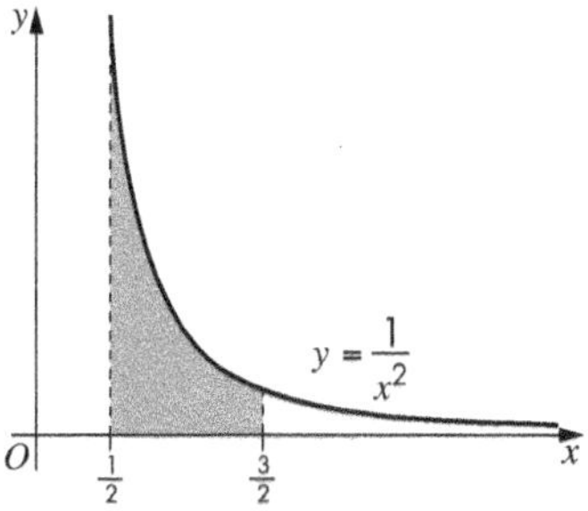

d

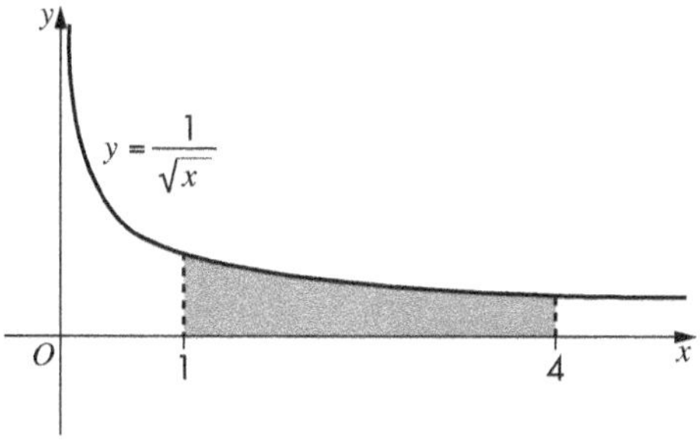

7 Find the area enclosed by the x-axis and each of these curves:

a $y = 3 - 3x^2$ **b** $y = 3x^2 - 6x$ **c** $y = x^2 - 2x - 3$ **d** $y = 2 - x - x^2$

8 Find the area enclosed by each curve and the x-axis.

a

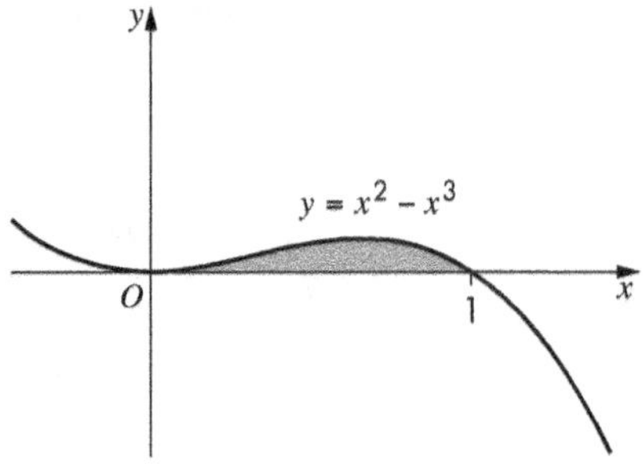

b

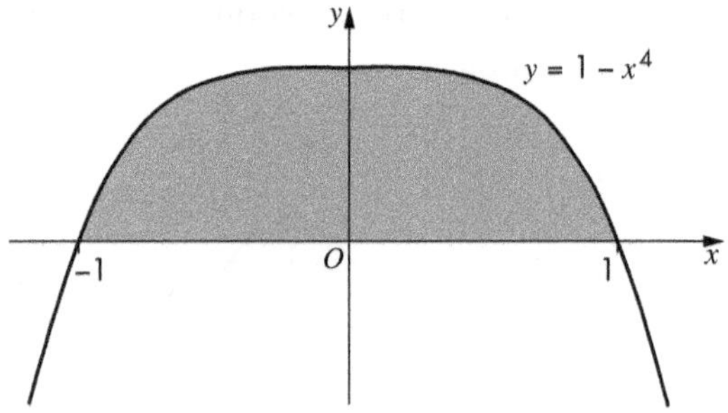

c

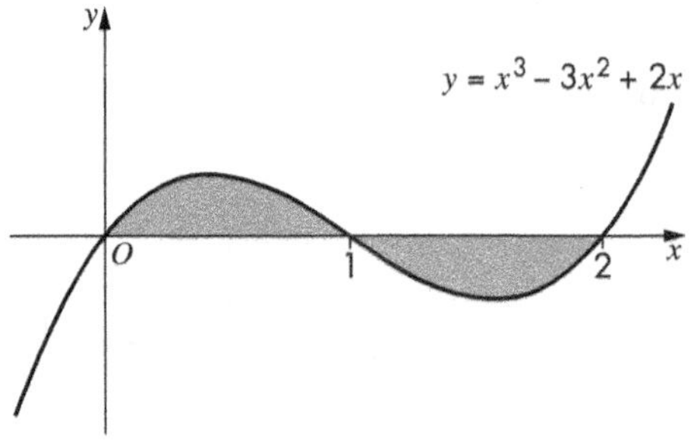

d

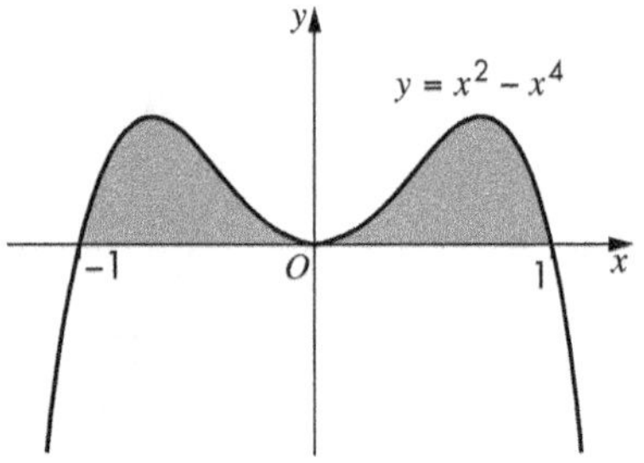

Exercise 17.2 Area between a trigonometric curve and the x-axis

Reminder

$\int \sin x \, dx = -\cos x + c \qquad \int \sin 2x \, dx = \dfrac{-\cos 2x}{2} + c$

$\int \cos x \, dx = \sin x + c \qquad \int \cos 2x \, dx = \dfrac{\sin 2x}{2} + c$

1 Calculate the exact value of each shaded area.

a

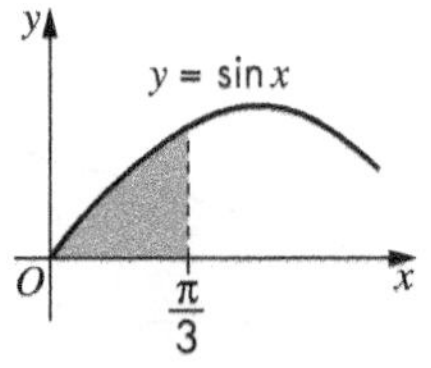

b

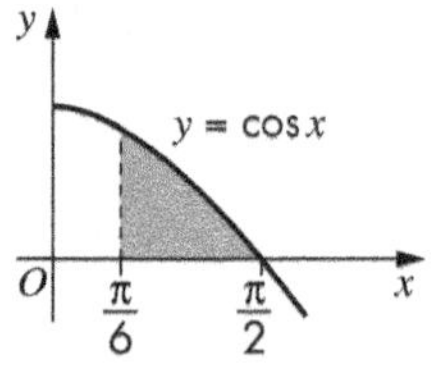

c

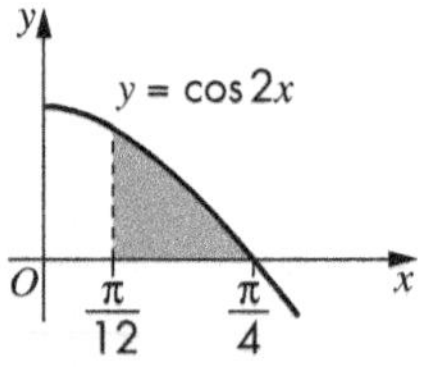

d

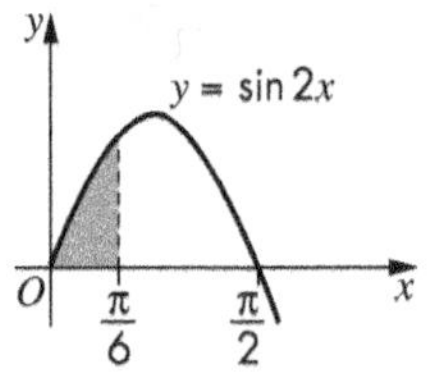

e

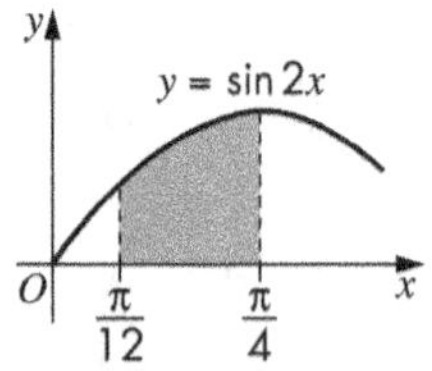

f

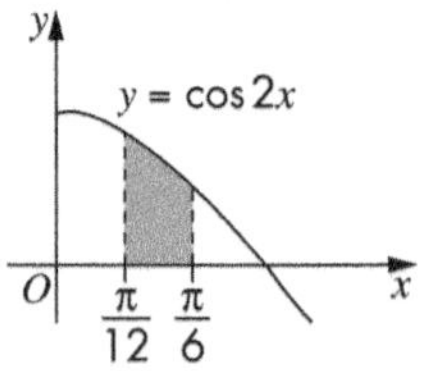

2 Calculate the shaded areas.

a

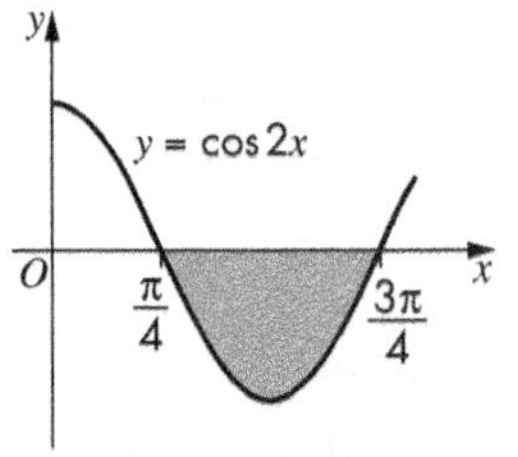

b

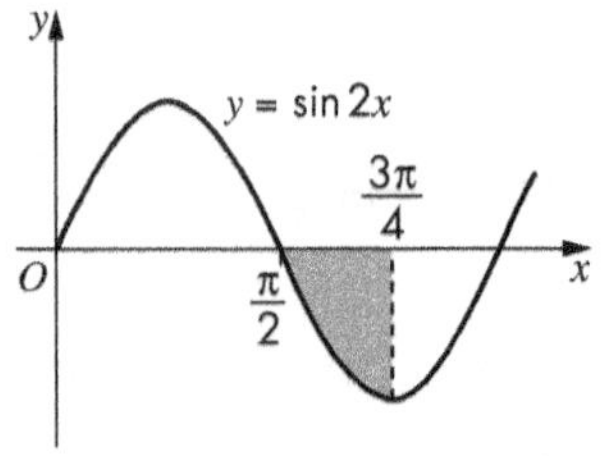

3

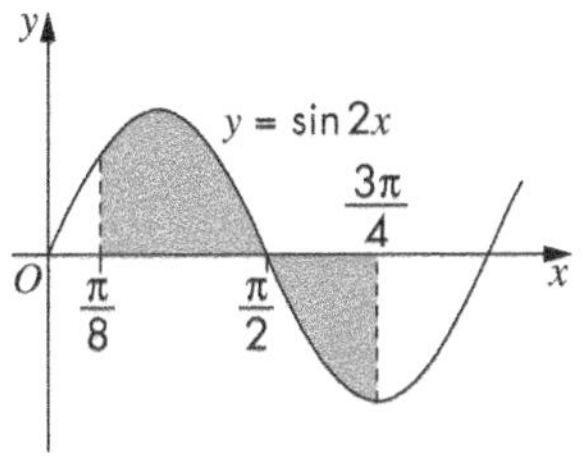

A design for a sand-timer is based on the shaded area in the diagram on the left.

Calculate the area of the shape used for the sand-timer.

4

a Calculate the exact value of shaded Area I shown in the diagram.

b Similarly calculate Area II.

c Show that the ratio of Area I to Area II is 9 : 1

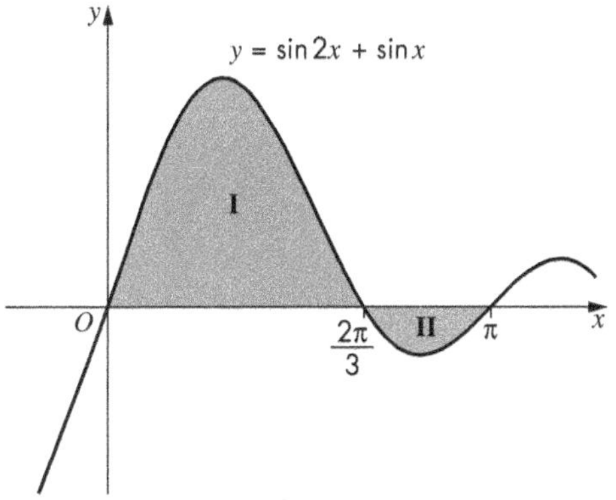

5

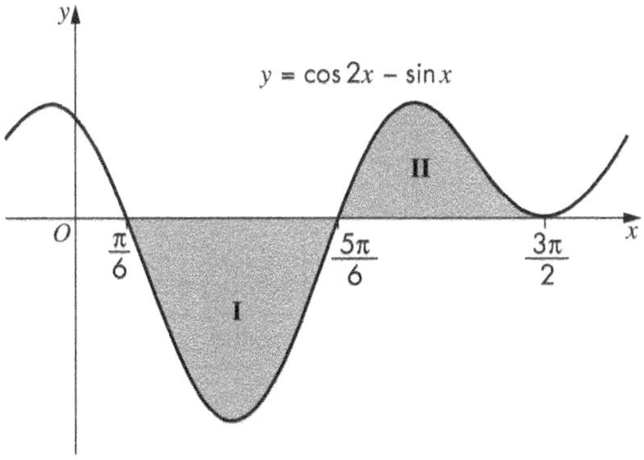

a Show that Area I is exactly $\frac{3\sqrt{3}}{2}$ units^2

b Prove that Area II is exactly half the area of Area I.

6

The diagram shows the graph $y = \cos 5x - \cos x$.

The graph has $x = \pi$ as an axis of symmetry.

The graph also has half-turn symmetry around the point $\left(\frac{\pi}{2}, 0\right)$.

Between 0 and 2π the graph intersects the x-axis at 7 points.

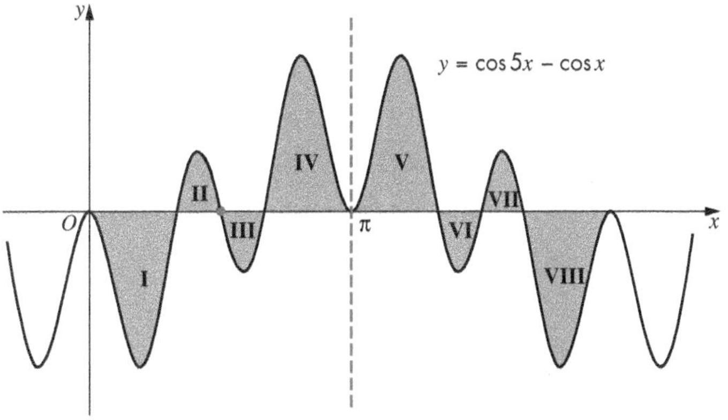

The first two points of intersection are $\left(\frac{\pi}{3}, 0\right)$ and $\left(\frac{\pi}{2}, 0\right)$. Use this information to answer the following.

a Find the other 5 points of intersection of the graph with the x-axis between 0 and 2π.

b Which areas are equal to Area I?

c Which areas are equal to Area II?

d Calculate the exact value of Area I.

e Calculate the exact value of Area II. Write your answer as a single fraction.

f Calculate Areas III – VIII using definite integrals.

Exercise 17.3 Area between two curves

Example

Find the area enclosed by $y = x^2 - 2x + 1$ and $y = x + 1$.

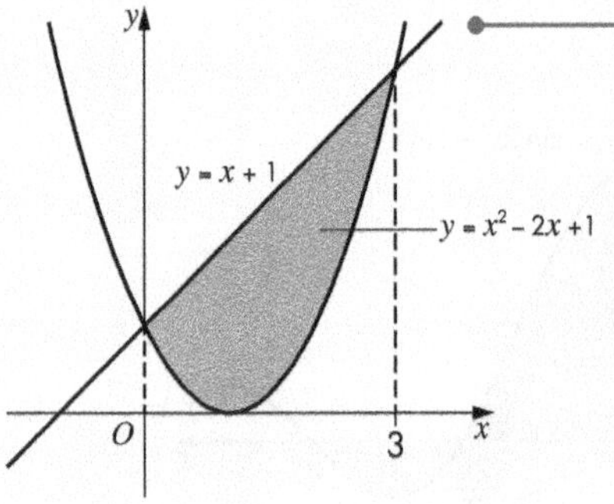

A sketch is essential.

The points of intersection can be found by setting the two equations equal and then solving: $x^2 - 2x + 1 = x + 1 \Rightarrow x^2 - 3x = 0 \Rightarrow x(x - 3) = 0$ giving $x = 0$ and $x = 3$.

Use $\int_a^b \left(\left(\begin{matrix}\textit{top}\\ \textit{graph}\end{matrix}\right) - \left(\begin{matrix}\textit{bottom}\\ \textit{graph}\end{matrix}\right)\right) dx$ with $x = a$ and $x = b$ being where the graphs intersect.

$\int_0^3 ((x + 1) - (x^2 - 2x + 1))\,dx$ — Always simplify **before** integrating.

$= \int_0^3 ((x + 1 - x^2 + 2x - 1))\,dx$ — Take care with the –ve sign outside brackets.

$= \int_0^3 (3x - x^2)dx$ — Use $\int ax^n\,dx = \dfrac{ax^{n+1}}{n+1} + c$ (but leave out c).

$= \left[\dfrac{3x^2}{2} - \dfrac{x^3}{3}\right]_0^3$ — Evaluate at $x = 3$ and evaluate at $x = 0$.

$= \left(\dfrac{3 \times 3^2}{2} - \dfrac{3^3}{3}\right) - \left(\dfrac{3 \times 0^2}{2} - \dfrac{0^3}{3}\right)$ — Practise calculating fractions without a calculator.

$= \dfrac{27}{2} - 9 - (0 - 0) = \dfrac{9}{2}$ — Value of integral in this case gives the area.

The required area is $\dfrac{9}{2}$ units^2

1 In each case:

i count squares to find the area of the shaded rectangle

ii construct a definite integral for determining the shaded area

iii evaluate the integral. Is the result the same as part **i**?

a

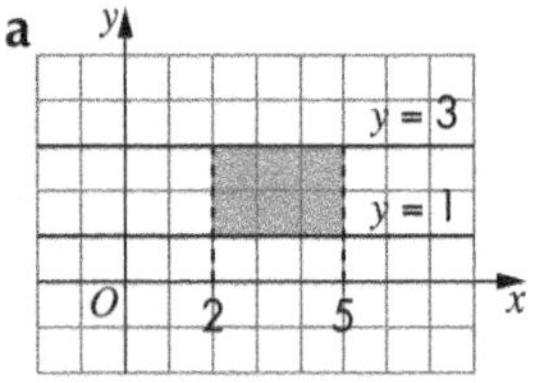

b

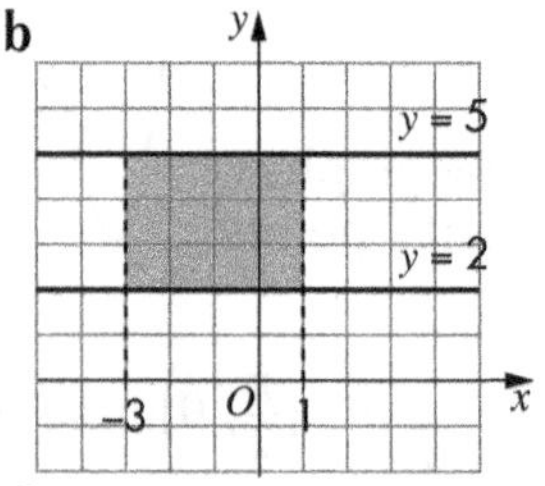

c

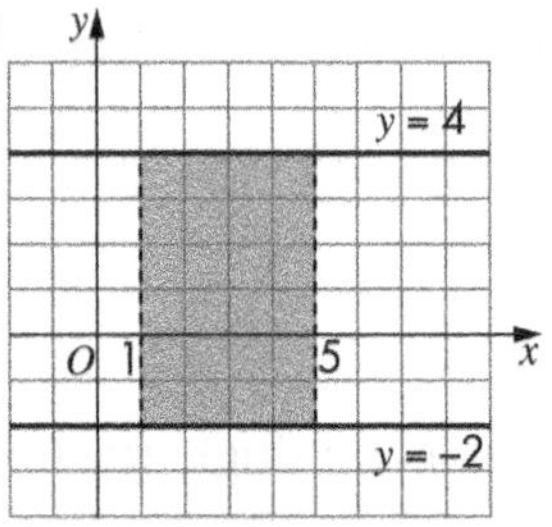

d

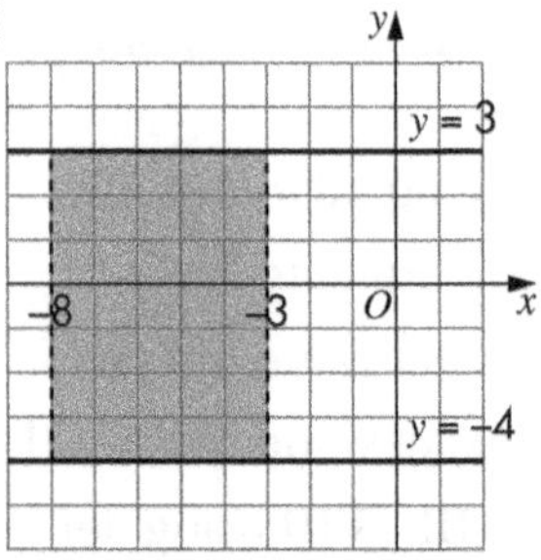

2 In each case:

i count squares to find the area of the shaded traingle

ii construct a definite integral for determining the shaded area

iii evaluate the integral. Is the result the same as part **i**?

a

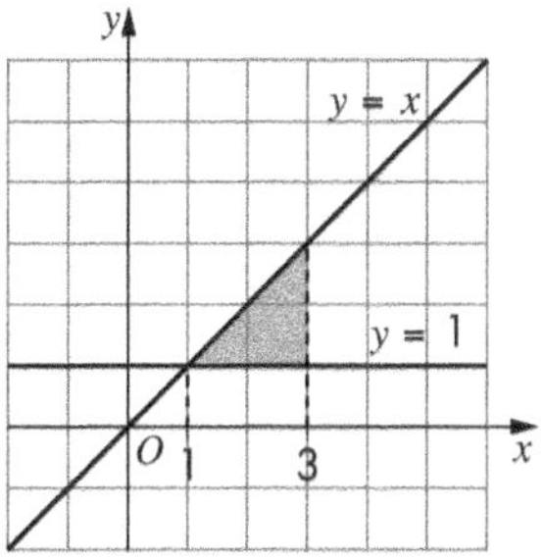

b

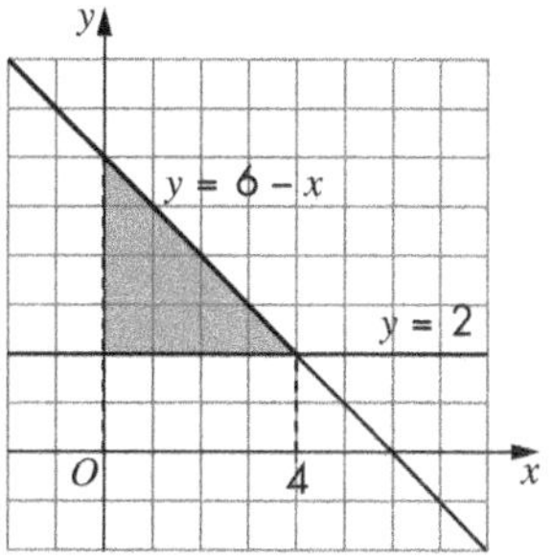

c

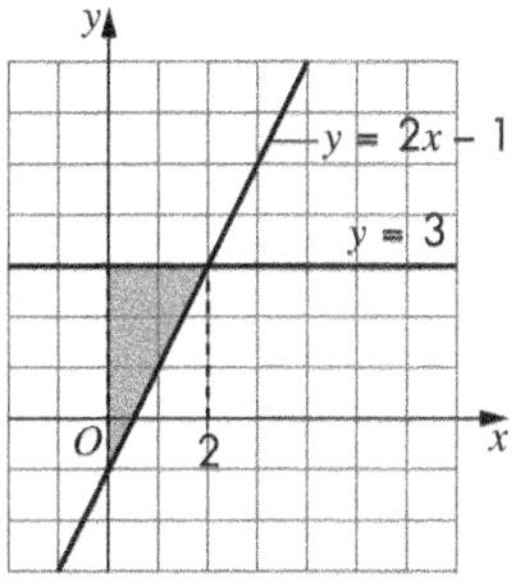

d

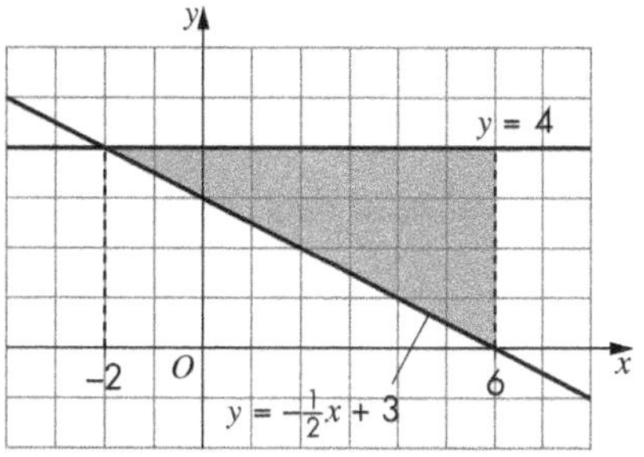

3 **a** For each of Area **I**, Area **II** and Area **III**, construct a definite integral and calculate the exact area.

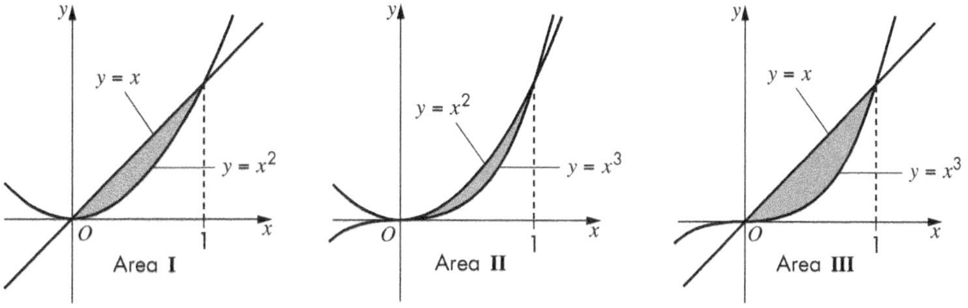

b What is the relationship between these three areas?

(Think of all three graphs on one diagram.)

4 Find the shaded area.

a

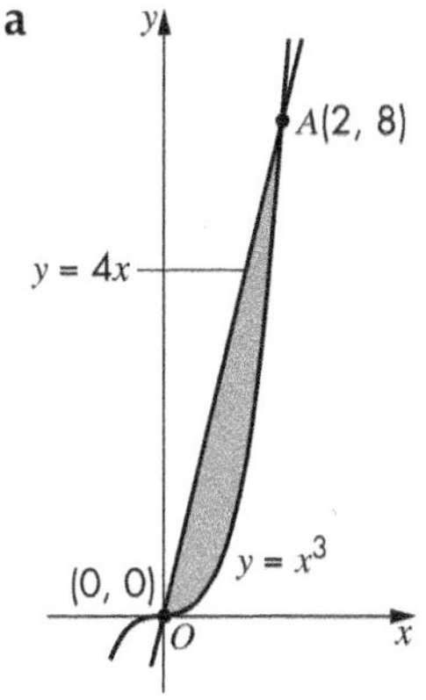

b

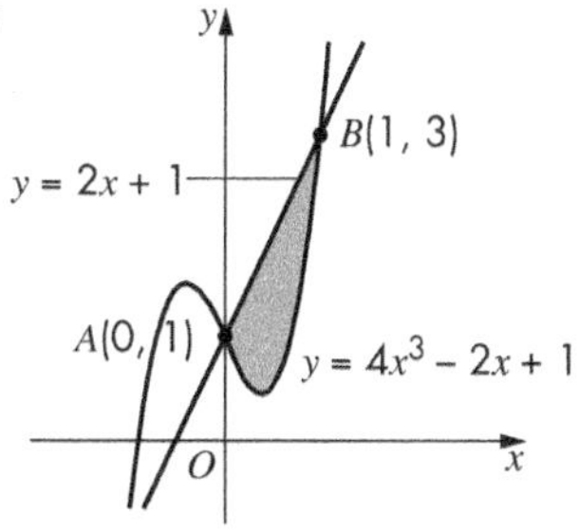

c

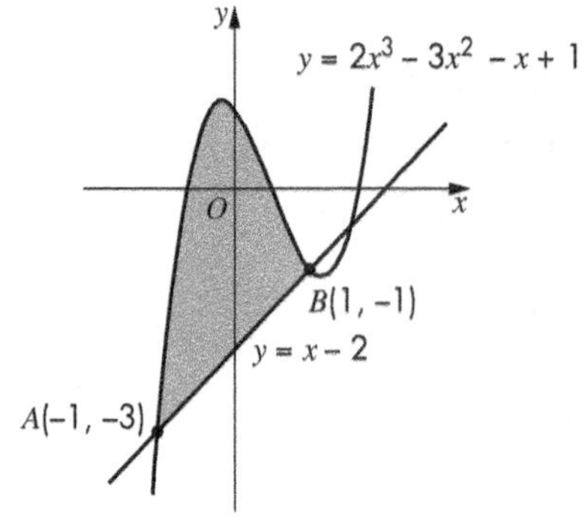

d

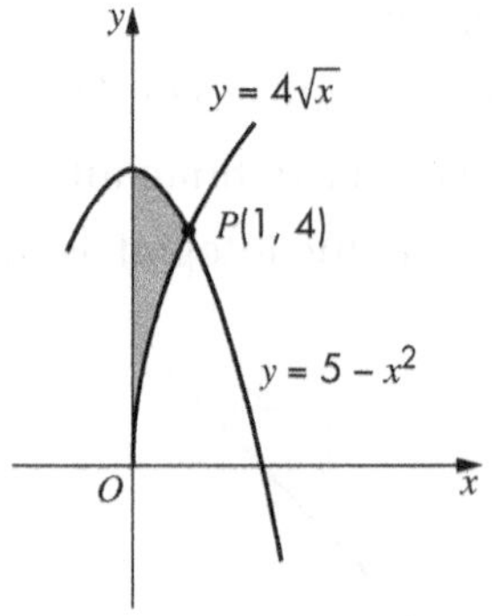

5 Find the area enclosed by these lines and curves:

a $y = x^2$
$y = x + 2$

b $y = 2x + 4$
$y = 4 - x^2$

c $y = 3x^2$
$y = 9x - 6$

d $y = \sqrt{x}$
$y = \frac{1}{2}x$

e $y = (x - 2)^2$
$y = -2x + 4$

f $y = 9 - x^2$
$y = x + 3$

6 Find the exact value of each of these shaded areas:

a

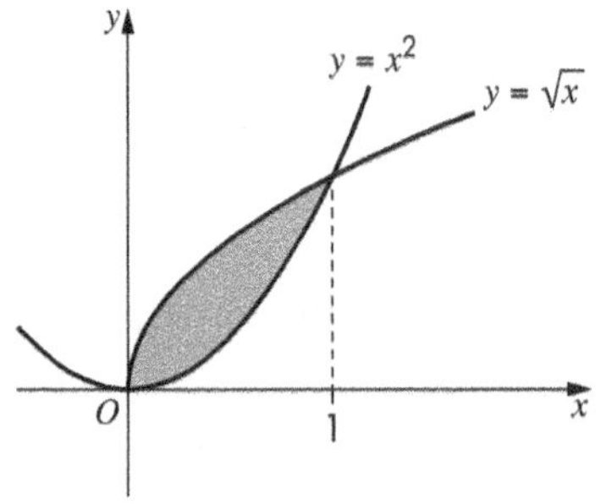

b

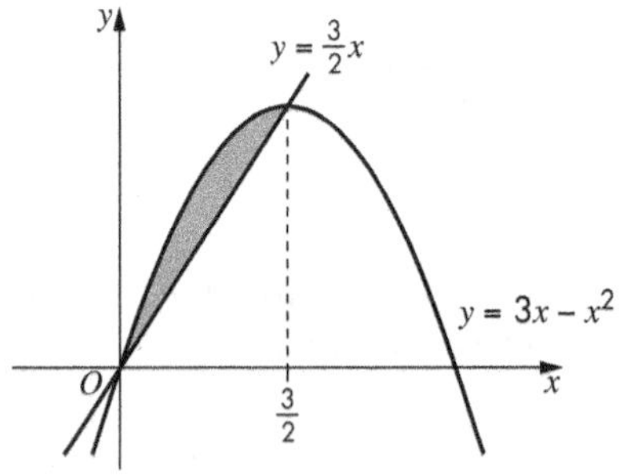

7 **a** Find the x-coordinates of the intersection points P, Q and R by solving the equation $\cos 2x = \sin x$.

b Construct definite integrals corresponding to Area I and to Area II.

c Evaluate these integrals to find the exact values of Area I and Area II.

d Find the ratio Area I : Area II

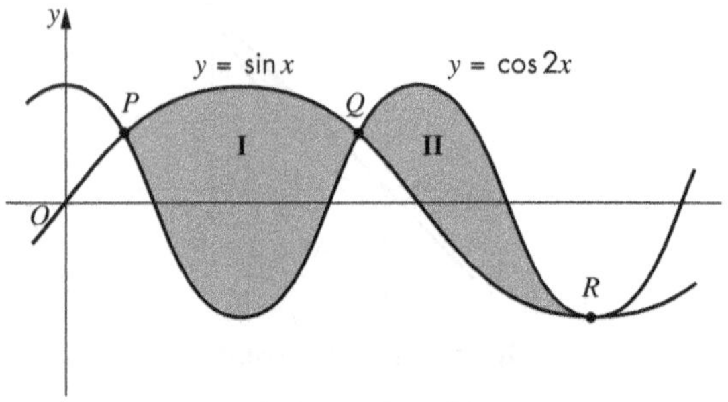

8 **a** Find the exact values of Area I, Area II, Area III and Area IV.

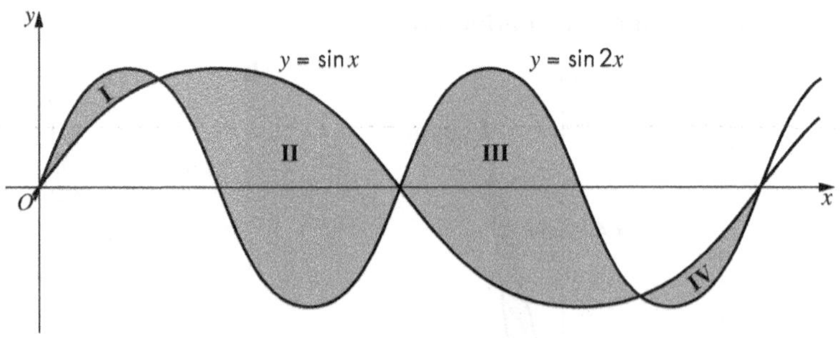

b Find the total shaded area.

c Find the ratio Area II : Area I

Notes

Notes

Notes

Notes